T0260303

Natural Language Processing and Information Retrieval

This book presents the basics and recent advancements in natural language processing and information retrieval in a single volume. It will serve as an ideal reference text for graduate students and academic researchers in interdisciplinary areas of electrical engineering, electronics engineering, computer engineering, and information technology. This text emphasizes the existing problem domains and possible new directions in natural language processing and information retrieval. It discusses the importance of information retrieval with the integration of machine learning, deep learning, and word embedding. This approach supports the quick evaluation of real-time data. It covers important topics including rumor detection techniques, sentiment analysis using graph-based techniques, social media data analysis, and language-independent text mining.

Features:

- Covers aspects of information retrieval in different areas including healthcare, data analysis, and machine translation
- Discusses recent advancements in language- and domain-independent information extraction from textual and/or multimodal data
- Explains models including decision making, random walks, knowledge graphs, word embedding, n-grams, and frequent pattern mining
- Provides integrated approaches to machine learning, deep learning, and word embedding for natural language processing
- Covers the latest datasets for natural language processing and information retrieval for social media like Twitter

The text is primarily written for graduate students and academic researchers in interdisciplinary areas of electrical engineering, electronics engineering, computer engineering, and information technology.

Computational and Intelligent Systems Series

In todays' world, the systems that integrate intelligence into machine-based applications are know as intelligent systems. In order to simplify the man-machine interaction, intelligent systems play an important role. The books under the proposed series will explain the fundamentals of intelligent systems, reviews the computational techniques and also offers step-by-step solutions of the practical problems. Aimed at senior undergraduate students, graduate students, academic researchers and professionals, the proposed series will focus on broad topics including artificial intelligence, deep learning, iImage processing, cyber physical systems, wireless security, mechatronics, cognitive computing, and industry 4.0.

Application of Soft Computing Techniques in Mechanical Engineering
Amar Patnaik, Vikas Kukshal, Pankaj Agarwal, Ankush Sharma, and Mahavir Choudhary

Computational Intelligence based Optimization of Manufacturing Process for Sustainable Materials
Deepak Sinwar, Kamalakanta Muduli, Vijaypal Singh Dhaka, and Vijander Singh

Advanced Mathematical Techniques in Computational and Intelligent Systems
Sandeep Singh, Aliakbar Montazer Haghighi, and Sandeep Dalal

Natural Language Processing and Information Retrieval
Principles and Applications

Edited by
Muskan Garg, Sandeep Kumar, and
Abdul Khader Jilani Saudagar

CRC Press
Taylor & Francis Group
Boca Raton London New York

CRC Press is an imprint of the
Taylor & Francis Group, an **informa** business

First edition published 2024
by CRC Press
6000 Broken Sound Parkway NW, Suite 300, Boca Raton, FL 33487-2742

and by CRC Press
4 Park Square, Milton Park, Abingdon, Oxon, OX14 4RN

CRC Press is an imprint of Taylor & Francis Group, LLC

ISBN: 978-1-032-15492-3 (hbk)
ISBN: 978-1-032-15493-0 (pbk)
ISBN: 978-1-003-24433-2 (ebk)

DOI: 10.1201/9781003244332

Typeset in Sabon
by codeMantra

Contents

Preface

Natural language processing supports a wide range of applications for healthcare reports, legal documents, automatic chat boxes, trend and event detection, machine translation, and multilingualism. The significant challenges of using word co-occurrence networks are text-normalization, collocation techniques for text segmentation, and integration of machine learning and deep learning approaches with graph-based techniques. Natural language processing is part of smart devices and computers like auto-complete suggestions for predictive typing while searching on Google, spell checkers, spam detection, etc. One of the most promising applications of natural language processing is efficiently handling users' queries on Google or elsewhere. Question-answering is an important research area for different industries for cost-cutting over the customer support team.

The challenges of user-generated and language/domain-independent text have been minimally explored. In addition, a huge amount of data is more suitable for statistical evaluation than semantic evaluation. To handle these issues, information retrieval has evolved for better natural language processing. This book promotes information retrieval by integrating machine learning, deep learning, and word embedding. There are a total of fourteen chapters; the first chapter discusses federated learning and its applications. The second chapter presents a near-real-time recommendation system based on utility itemset mining by employing the highly efficient EAHUIM (Enhanced Absolute High Utility Itemset Mining) algorithm, which has been proposed for finding the associations between items in a database. Also, as the real-world data is dynamic, the problem of transaction-stream fluctuation has been considered, and an adaptive load strategy has been proposed for the model. Various experiments on a real-world dataset result in near-real-time and customized recommendations, demonstrating the model's efficacy. The next two chapters discuss anaphora resolution, neural machine translation, and multi-document extractive summarization.

Chapter 4 discusses the processing of questions, documents, and answers. A deep neural network for question-answering systems was implemented by the authors. Chapter 5 presents the evolution of search and landmark events in the history of search engines. Next, different dimensions of "semantic analysis" research fields are discussed. Then, the catalyst events for research progress in the contextual research domains, specifically automatic question-answering, are described. Eventually, automatic question-answering proved to be the most promising and discrete field that evolved from search engines.

Chapter 6 considers code-switching for study. Code-switching is typically considered a feature of informal communication, such as casual speech or online texting. However, there is a plethora of evidence that language alteration also happens in formal situations, for instance, in newspaper headlines, political speeches, and teaching. A hybrid model is proposed in the next chapter for legal document summarization. Chapter 8 illustrates network-text analysis. A network-text analysis is a way to extract knowledge from texts and generate a network of words. A central premise is that the network represents a mental model of the author. After transforming an unstructured text into a structured network, text analytic methods can be used to analyze the network conducted by specific networks.

Chapter 9 focuses on machine reading comprehension (MRC). This chapter incorporates the MRC architecture, modules, tasks, benchmarked datasets, and performance evaluation metrics. A new classification of the MRC tasks was included with representative datasets. Furthermore, the deep neural network, Transformer, BERT, and BERT-based models are explored. Chapter 10 explores the necessity of question-answering system in the current education system. The authors propose a question-answering system for subjective questions.

Editors

Dr. Sandeep Kumar is currently a professor at CHRIST (Deemed to be University), Bangalore. He recently completed his postdoctoral research at Imam Mohammad Ibn Saud Islamic University, Riyadh, Saudi Arabia, in sentiment analysis. He is an associate editor for Springer's *Human-Centric Computing and Information Sciences* (HCIS) journal. He has published more than eighty research papers in various international journals/conferences and attended several national and international conferences and workshops. He has authored/edited seven books in the area of computer science. Also, he has been serving as General Chair of the International Conference on Communication and Computational Technologies (ICCCT 2021, 22, and 23) and the Congress on Intelligent Systems (CIS 2022 and 2023). His research interests include nature-inspired algorithms, swarm intelligence, soft computing, and computational intelligence.

Dr. Abdul Khader Jilani Saudagar received the Bachelor of Engineering (B.E.), Master of Technology (M.Tech.), and Doctor of Philosophy (Ph.D.) degrees in computer science and engineering in 2001, 2006, and 2010, respectively. He is currently working as an Associate Professor with the Information Systems Department, College of Computer and Information Sciences (CCIS), Imam Mohammad Ibn Saud Islamic University (IMSIU), Riyadh, Saudi Arabia. He is also the Head of the Intelligent Interactive Systems Research Group (IISRG), CCIS. He has ten years of research and teaching experience at both the undergraduate (UG) and postgraduate (PG) levels. He was the Principal Investigator of the funded projects from KACST, the Deanship of Scientific Research (IMSIU), and is working as the Principal Investigator for the project titled "Usage of modern technologies to predict emergence of infectious diseases and to detect outbreak of pandemics" in the grand challenge track, funded by the Ministry of Education, Saudi Arabia. He has published a number of research papers in international journals and conferences. His research interests include artificial image processing, information technology, databases, and web and mobile application development.

He is associated as a member with various professional bodies, such as ACM, IACSIT, IAENG, and ISTE. He is working as an editorial board member and a reviewer for many international journals.

Dr. Muskan Garg is working as a Postdoctoral Research Fellow at Mayo Clinic, Rochester, Minnesota. Prior to Mayo Clinic, she served as Postdoctoral Research Associate at the University of Florida. Prior to UFL, she served as an Assistant Professor at Computer Science and Engineering Department at Thapar Institute of Engineering & Technology, India. She has completed her master's and doctorate degrees from Panjab University, Chandigarh. Her previous research work is associated with applied network science in the fields of Information Retrieval and Natural Language Processing for social media data. In general, she is interested in exploring the domains of causal analysis, semantic relations, ethics, and social media data, as well as mental health analysis on social media.

Contributors

Farzana Bhuiyan
University of Chittagong
Chittagong 4331, Bangladesh

Md Masum Billah
American International University Bangladesh (AIUB)
Kuratoli 1229, Bangladesh

Preetpal Kaur Buttar
Department of Computer Science and Engineering
Sant Longowal Institute of Engineering and Technology
Longowal, Sangrur, Punjab, India

Vandna Dahiya
Department of Computer Science and Applications
Maharshi Dayanand University, Rohtak, Haryana, India

Arijit Das
Department of CSE
Faculty of Engineering and Technology
Jadavpur University
Jadavpur, Kolkata, West Bengal 700032, India

Nandhini K
Department of Computer Science
School of Mathematics and Computer Sciences
Central University of Tamil Nadu
Thiruvarur, Tamil Nadu 610005, India

Madhav A. Kankhar
Department of Computer Science & Information Technology
Dr. Babasaheb Ambedkar Marathwada University
Aurangabad (MS), India

Mohammed Kaosar
Murdoch University
Murdoch, WA 6150, Australia

Kalpana B. Khandale
Department of Computer Science & I.T.
Dr. Babasaheb Ambedkar Marathwada University
Aurangabad (MS), India

C. Namrata Mahender
Department of Computer Science & I.T.
Dr. Babasaheb Ambedkar Marathwada University
Aurangabad (MS), India

Anastasia Nikiforova
Sberbank
Russia

M. Punithavalli
Department of Computer Applications
Bharathiar University
Coimbatore, Tamil Nadu 641046, India

Manoj Kumar Sachan
Department of Computer Science and Engineering
Sant Longowal Institute of Engineering and Technology
Longowal, Sangrur, Punjab 148106, India

Diganta Saha
Department of CSE
Faculty of Engineering and Technology
Jadavpur University
Jadavpur, Kolkata, West Bengal 700032, India

Dipanita Saha
Noakhali Science and Technology University
Noakhali 3814, Bangladesh

Bharat A. Shelke
Department of Computer Science & Information Technology
Dr. Babasaheb Ambedkar Marathwada University
Aurangabad (MS), India

Sergei Ternovykh
Rostelecom Contact Center
Russia

Nisha Varghese
Department of Computer Applications
Christ University
Bangalore, Karnataka 560035, India

Deekshitha
Department of Computer Science
School of Mathematics and Computer Sciences
Central University of Tamil Nadu
Thiruvarur, Tamil Nadu 610005, India

Chapter 1

Federated learning for natural language processing

Sergei Ternovykh
Rostelecom Contact Center

Anastasia Nikiforova
Sberbank

1.1 INTRODUCTION

Over the past few years, the legislation of many countries (including the USA, European Union, and China) has been updated with provisions prohibiting the transfer of personal data (medical and biological data) of their residents outside the country. Besides, sometimes it is unclear who is the owner of the data. For example, a call center that receives calls for the benefit of its customers keeps calling records. However, does it have the right to provide them to anyone? Customers who order and pay for campaigns may also be the data owners. Moreover, some rights may belong to people who call the call center and whose voices are heard on the recordings.

Training quality for machine learning models can be improved for some tasks by using the data people keep on their mobile devices. For example, a next typing word prediction model provides users with more relevant suggestions. A model for assessing photo quality can tell them which of their photos can be deleted and which are worth keeping. The first task requires access to user chat history; the second requires access to the photo archive and the editing history. All these data are private to the users and cannot be downloaded from their devices. Even de-identified data may accidentally reveal some sensitive information [1].

However, it is possible to arrange training without copying the data. Algorithms united by the federated learning (FL) paradigm are used to solve this problem. As defined in [2], *Federated Learning is a machine learning setting where multiple entities (clients) collaborate in solving a machine learning problem under the coordination of a central server or service provider. Each client's raw data is stored locally and not exchanged or transferred; instead, focused updates intended for immediate aggregation are used to achieve the learning objective.*

Although researchers have been interested in privacy-preserved data analysis for several decades, only wide usage of the Internet and machine

DOI: 10.1201/9781003244332-1

learning allowed the issue of using privacy-preserved data in training to become relevant. In 2010, an approach was presented based on the differential privacy [3] and secure multiparty computation [4–6] concepts, within which it was proposed to aggregate classifiers independently trained on individual data instances into a single model [7]. However, the resulting models did not continually improve.

In 2015, the distributed selective SGD (DSSGD) algorithm was proposed [8], allowing multiple participants to train identical neural network models on their data without sharing them. They only exchange model parameters through the server during training, while the clients are trained sequentially or in asynchronous mode. In a laboratory setting, this method has shown promising results, and it seems that it could be used to train models with the joint efforts of several organizations. However, the algorithm did not consider the heterogeneity of the data.

Almost simultaneously with DSSGD, a paradigm, at first called by the authors *Federated Optimization*, was formulated in [9]. It took a similar approach, but the goal was to train a model on data distributed across multiple mobile devices. The data itself during training also remained on the devices without sharing. Clients were trained in parallel at each round, and the server was responsible for synchronizing the learning outcomes. The modern name – **Federated Learning** (here, *federated* means a *learning task is solved by a loose federation of participating devices*) – was first used for this technique by Google in [10]. Also, this work proposed the still widely used algorithm FedAvg for aggregating client updates.

Since privacy-preserved data can be located not only on mobile devices, the concept of FL covers any methods of collaborative training of models when data does not leave the infrastructure of the owners.

Situations where the need to ensure privacy excludes the possibility of centralized learning occur in many domains. *Natural language processing* (NLP) is just one of them, and there are no noteworthy differences in the usage of privacy-protected learning algorithms for NLP tasks. Practical examples of applying FL to NLP show that there are still unresolved problems in achieving optimal model performance. However, the basic concepts and methods considered below are domain-independent, and the choice of any of them is mainly determined by the mutual trust of those involved in solving the particular task and the technical capabilities of their hardware.

1.1.1   Centralized (standard) federated learning

FL originally implied a central server to coordinate the training process. With this approach, the learning setting is built as a hub-and-spoke topology (Figure 1.1), and, in the general case, interactions are carried out according to the following workflow:

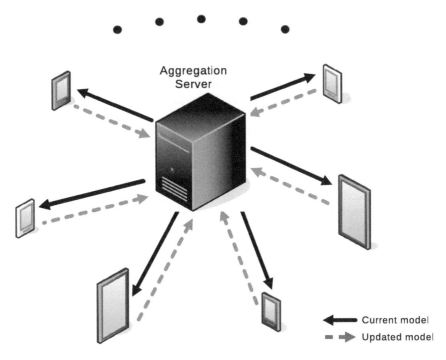

Figure 1.1 An example of federated learning architecture: hub-and-spoke topology.

1. From the clients that satisfy the predetermined conditions, the server selects a subset of a specific size and sends them the current version of the model.
2. Clients train the received model on their local data for a specified number of epochs and send updates of the model parameters or the updated parameters to the server.
3. The server aggregates the training results and produces an updated model version.
4. The cycle is repeated until it converges.

It is worth noting that although all clients participate in training the same model, the training hyperparameters (learning rate, number of epochs, objective function, etc.) in the general case may vary across clients.

This technology has been studied quite well. It has flaws and unresolved issues (see below), but they are mostly known. If the project participants accept their presence, it is easy to organize the learning process. There are tools for building workflows in the "sandbox" and launching projects in production environments.

1.1.2 Aggregation algorithms

Algorithms that aggregate client updates directly affect the performance of the trained model. One of the main challenges in FL is the unbalance of datasets. Often, the data is not identically or independently distributed across clients. In addition, the amount of data on two clients from one learning setting may differ by orders of magnitude. The algorithms used in FL should take into account these aspects.

Another critical factor primarily affects the wall-clock time of learning. It is the amount of data transmitted over the network. At each training round, the server sends the current slice of the model parameters to selected clients, and after the round has been completed, it receives the new parameters (or their updates). It can be hundreds of megabytes of data, so a slow transmission channel may significantly increase the overall model convergence time. Various compression and quantization methods are used to reduce the influence of this factor; however, the main role in the optimization of network exchange is played by aggregation algorithms.

Now, let us consider the main features of some of the most recent proposed algorithms in more detail.

1.1.3 DSSGD

DSSGD [8] assumes that not all model parameters equally affect its performance. Therefore, in each round, participants exchange only those parameters that were marked as the most relevant during the training.

At the beginning of the training, the server sends all clients equally initialized model instances. During the training round, clients either sequentially (with round-robin or random order) or in asynchronous mode download from the server the model parameters that need to be changed, copy them into their model, and run a training step. After that, they send the following to the server:

- *Largest values*: Top-N gradients, largest in absolute value.
- *Random with threshold*: Randomly subsample the gradients whose value is above threshold τ.

The server applies gradients uploaded by the latest client to its model and sends the next client the most frequently updated parameters during all previous training rounds. The client copies these parameters into its model instance, continuing the process.

The authors show that cooperation, even sharing only 1% of parameters, results in higher model performance than standalone learning on any client's datasets. DSSGD with round-robin client selection performs almost as well as centralized training when all the data are pooled into one dataset and the model is trained locally using standard SGD. Moreover, the approach provides the required level of data privacy, and the amount of data transmitted over the network is relatively small.

At the end of the training, each participant has a personalized version of the trained model (i.e., fine-tuned for this exact client).

The sequential nature of processing the client updates prevents the use of the algorithm in settings with a high number of clients since the wall-clock time of the training, in this case, would be extremely high. However, [8] evaluates learning settings with 30, 90, and 150 nodes and shows that an increase in the number of nodes leads to an increase in the resulting model performance.

Unfortunately, the authors did not consider the statistical heterogeneity of actual client datasets in their experiments. They just used chunked MNIST and SVHN datasets, so the high results were not surprising.

1.1.4 FedAvg

FedAvg (FederatedAverage) [10] is still the most commonly used aggregation algorithm for clients' updates. Its authors notice the disadvantages of DSSGD and propose an alternate solution. As DSSGD, it aims to reduce the amount of information transmitted over the network but also take into account the heterogeneity of client data. Additionally, mobile devices are considered clients in that solution, so their number in the general case is exceptionally high.

As a baseline, the authors consider an algorithm called FedSGD (FederatedSGD), which works as follows. On each round, a C-fraction of clients is selected, and each client computes the loss over all the data it holds. Thus, C controls the global batch size, with C=1 corresponding to full-batch (non-stochastic) gradient descent. Then, the server receives clients' gradients and calculates a weighted sum (each client's weight equals the ratio of its local dataset size to the total size of all client datasets). After that, the server runs backpropagation.

The FedAvg algorithm is a generalization of FedSGD that adds an extra computing step to the client. Having received a loss, clients perform backpropagation by themselves through their local model instances and repeat the training cycle several times. After that, parameters of the client model instances are sent to the server, which carries out their weighted averaging. Thus, the algorithm is described by three parameters:

- C – the fraction of clients selecting on each round;
- E – the number of training passes (epochs) each client makes over its local dataset on each round;
- B – the local minibatch size used for the client updates.

With $E=1$ and $B=\infty$, FedAvg is equal to FedSGD.

Experimental results show the high efficiency of the solution. With correct E and B hyperparameter selection, FedAvg significantly outperforms the FedSGD baseline in the number of rounds for divergence and the resulting model accuracy. Choosing the right learning rate is also very important. If all the hyperparameters are optimal, FedAvg has comparable performance to centralized learning for the case when clients' data is **IID** (identically and independently distributed) [11].

It is worth noting that high performance is achieved when all clients at the beginning of training receive model instances with identically initialized parameters.

1.1.5 FedProx

Although FedAvg has become the de facto standard method for FL, it does not fully address the complexities of heterogeneity. The algorithm simply drops weak devices that do not manage to complete E training epochs in a given time. Besides, the algorithm does not always converge when data is not identically distributed across devices in some conditions.

To address these challenges, [12] proposed a novel algorithm called FedProx that can be considered a parametrized FedAvg.

In FedAvg, choosing the right hyperparameters is essential. A low value of the number of epochs E leads to a significant increase in network traffic. However, with a high value of this hyperparameter, model instances on some devices may be overfitted, i.e., locked in local minima.

FedProx tolerates the requirement of constant work for all devices by adding to the objective function that in FedAvg is equal for all the clients, special proximal term to prevent clients' local model overfitting. As a result, the number of local epochs may vary across clients and rounds.

FedProx does not bring any other additions to the FedAvg algorithm. So, if the proximal term is set to 0, FedProx is the same as FedAvg. However, in device and dataset heterogeneity, this approach results in more robust and stable convergence than vanilla FedAvg. Experiments show that the absolute testing accuracy is improved by 22% on average in highly heterogeneous settings.

1.1.6 SCAFFOLD

The authors of [13] examined the performance of FedAvg and concluded that when data is heterogeneous, FedAvg suffers from *client drift*. In particular, during the training process, the update direction for each client model instance differs from the resulting update direction of the server model (each client model is biased toward its data). This can result in slow and sometimes unguaranteed convergence.

To correct this, [13] proposes a **SCAFFOLD** ("Stochastic Controlled Averaging") algorithm that reduces variances in local updates. SCAFFOLD

adds to the FedAvg additional variables, called control variates, that estimate the update directions for the server and each client model instance.
Initially, all these variables had zero values. After backpropagation, each
client participating in the current round calculates its control variate. There
are two versions of it: in the first, it is just a gradient. Then, the client sends
a delta between its current and previous control variates to the server, and
the server adds the mean of all received clients' values to its control variate.
In the next step, the server sends this control variate to clients, and they use
the difference between this value and their previous control variates to correct their gradients after each local training pass.

As a result, the algorithm requires significantly fewer communication
rounds than FedAvg and is not affected by unbalanced client datasets.

The method tries to achieve the same goal as FedProx: do not allow client models to fall into local minima, i.e., prevent them from overfitting.
However, they use different approaches.

1.1.7 FedOpt

Another exciting generalization of FedAvg comes from Google. The authors
of [14] noted that FedAvg suffers from a lack of adaptivity besides *client
drift*. While FedAvg uses SGD for local optimization, it is unsuitable for
heavy-tailed stochastic gradient noise distributions, e.g., those that often
occur in models based on the transformer architecture [15].

The authors rethought the FedAvg architecture and noticed that the
server-side update equals the SGD step with a learning rate of 1.0. The
averaging step can be represented as applying averaged client gradients (or
pseudo-gradients) to the current server model.

The authors proposed replacing the averaging step with a natural optimization step of a particular method. They have implemented **FedOpt** versions for *AdaGrad*, *Adam*, and *Yogi* algorithms (**FedAdagrad**, **FedAdam**,
and **FedYogi**, respectively). In their experiments, they also tested *SGD with
momentum*.

Additionally, learning rate decay was added on the client side to prevent
client drift. It helped to reduce the gap between federated and centralized
learning, and the best performance was achieved with *ExpDecay*, where
the client learning rate is decreased by a factor of 0.1 every 500 rounds.

The experiments have shown that the new approach surpasses the vanilla
SGD in terms of both convergence time and the resulting model performance
on many tasks. For several FL tasks, new state-of-the-art results have been
achieved. As a by-product, comprehensive and reproducible empirical benchmarks for comparing federated optimization methods were introduced.

Also, the authors checked SCAFFOLD and stated that in their tests,
it performed comparable to or even worse than FedAvg. Citing [16], the
authors note that, while theoretically performant, the SCAFFOLD variance
reduction method may often perform worse than SGD.

1.1.8 Other algorithms

Above, we have described several important aggregation algorithms for FL. However, this list is not exhausted. Many interesting approaches have been proposed recently: **FedAtt** ("Attentive Federated Aggregation") [17], **FedNova** ("Federated Normalized Averaging") [18], **FedPD** ("Federated Primal-Dual") [19], **FedDyn** ("Federated Learning based on Dynamic Regularization") [20], etc. Nevertheless, the **FedAvg** method, proposed along with the FL paradigm, remains the most frequently used. Moreover, it is recommended to use FedAvg as a baseline when arranging a new FL project.

1.1.8.1 The FL categorization by client types

FL was initially developed and used by Google to train models on mobile users' data. In particular, the Gboard application implemented the method to improve next typing word suggestions [21]. Currently, such a setting type is called *cross-device*, in contrast to the *cross-silo* setting, where a relatively small number of organizations with large amounts of data cooperate to solve a problem [2].

1.1.9 The cross-device setting

This configuration, as mentioned above, corresponds to the case when clients are a considerable number of edge devices, each containing a relatively small amount of data (mobile phones, IoT devices, etc.). The learning setting, in this case, can contain billions of clients, and only a tiny subset of them is selected for each round. These devices can drop out of the network at any time (usually, 5% or more clients are expected to fail or drop out [2]), and learning algorithms must be aware of this factor.

This setting is often used when the organizer of the training is also the developer of an application that is already installed on user devices. In this case, the organizer does not need to obtain users' consent. It is assumed that, since private data are not transmitted outside and the training contributes to the improvement of the application they use, device owners a priori have no objection to participation. Of course, the corresponding clause in the license agreement would not be superfluous.

The learning process should not create problems for device owners. The technology authors determined from the beginning [2] that the clients participating in the following training round should be charged, plugged in, and on an unmetered wi-fi connection. Typically, devices meet these requirements at night. Furthermore, since local time depends on geographic location, we can violate client data's independence by using a geographical pattern. On the other side, if we try to select an equal number of devices from different time zones, then we can get biased in favor of users whose devices are connected to the power grid during periods when the majority of the population in their area carries phones with them [11].

A similar challenge arises due to the unbalanced sizes of datasets on devices. The more training data on a particular device, the greater the effect of the output from that device on the resulting model [2]. Suppose active users of our application are located in a specific territory or belong to a particular social group. In that case, we either get biased in favor of these territories/social groups, or if we try to smooth the situation, then, as in the example above, we get biased in favor of people who use our application, even though it is unpopular in their environment. Regions/groups where the application is not used are unrepresented in any case.

Another important factor is device heterogeneity. For example, some devices may have less storage than is required to train our model. Before launching, we have to exclude such devices from the settings. However, even with enough memory, the lack of computational or communication capacities can become an obstacle [22]. We know that perhaps not all clients selected for the next round can complete their training processes, so we chose a number of them with a margin. Also, we set the maximum waiting time for responses. If we have received the required number of responses by the time the wait ends, the round is considered successful (in some scenarios, we stop waiting immediately after getting the required responses). However, the probability that we will receive a response from a particular device depends on its characteristics and the transmission medium to which it is connected. Due to this, our model may be biased toward higher-income users who can afford better devices and services [10].

Statistical heterogeneity is the main challenge that complicates model training in the FL paradigm [22]. The clients' data is often **non-IID** (not identically or independently distributed). This is an inherent problem for any learning technology on different datasets without direct access to them, and the setting we use does not matter.

Some model biases (for example, biases in favor of active users) may be positive factors. Also, the model bias can be, in a way, neutralized with the additional tuning of the model on users' devices before usage (*model personalization*). This can be done either on the local data of the particular client itself or on the data of a group of similar clients (*user clustering* [23]). A description of modern approaches to model personalization can be found in [22–24]. However, the problem of finding and eliminating bias in FL and ML at large is generally unsolved, and currently, its relevance is only increasing.

1.1.10 The cross-silo setting

In this case, clients are organizations or data centers that are relatively few (usually less than 100). At the same time, each has a significant amount of training data. Within such a setting, all clients can participate in each round, and with a reasonable degree of confidence, it can be assumed that none of them disconnects during the training process.

However, to arrange such a project, one needs to find an incentive for companies to participate. On the one hand, any particular company's quality of the models directly affects its operational efficiency. On the other hand, if a company already has a better model than its competitors, it only joins the project if there is a risk that the competitors, having united, may create a model that performs even better. At the same time, there are always reasonable doubts that all the participants provide comparable training data (due to lack of it or intentionally). Some may even contribute spam/false information to reduce the resulting model's performance [25]. If it is assumed that all participants obtain an identical model after training, then the company that hid the data could later use it for additional model tuning and gain an unfair advantage. Aside from this, although only owners are granted access to their data within the FL paradigm, the potential threat of data leakage through FL is relatively high (see below).

Clients refuse to participate in FL projects unless they expect to receive sufficient rewards. The challenge of finding effective incentive schemes for FL is now in the active research phase. There are two main challenges [26]: how to evaluate each client's contribution, and how to recruit and retain more clients.

The current state of this issue is summarized in [26,27]. Reference [26] classifies incentive mechanism designs proposed so far as driven by clients' *data contribution* (based on either *data quality* or *data quantity*), *reputation* (when participants rate each other), and *resource allocation* (considered *computational* and *communication* resource allocation). Reference [27] builds a taxonomy of state-of-the-art incentive mechanisms in a technical way, based on algorithmic and engineering implementation details.

However, the existing solutions are still not convincing enough. The task is complicated because some clients may participate in the project for malicious purposes. In this case, they can deliberately transfer model updates to degrade its performance. Finding methods to detect and ban/punish such clients is an important new research area that is just beginning to develop. As specified in [26,27], the incentive mechanism design for FL is in its infancy, and many open issues exist.

1.1.10.1 The FL categorization by data partitioning

In [28], a FL categorization based on the data's distribution characteristics was introduced.

1.1.11 Horizontal federated learning

The cross-device setting usually deals with distributed data that share the same feature space. Such data is referred to as *horizontally* or *sample/*

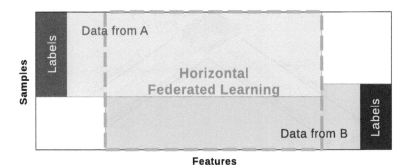

Figure 1.2 Horizontal (sample-partitioned) federated learning. (Based on [80].)

example partitioned, and a set of approaches aimed at processing such data in the FL paradigm, respectively, *Horizontal* or *Sample-based Federated Learning* (Figure 1.2).

1.1.12 Vertical federated learning

The cross-silo setting can be more flexible. Several organizations, especially those in the same domain, can store much of the same information about their customers. However, it is not uncommon when some companies from the same region have largely overlapping subsets of users, but some of the user data is unique to each company. Such data is called *vertically* or *feature partitioned*, and the privacy-preserving methods of its processing form the family of *Vertical* or *Feature-based Federated Learning* (Figure 1.3).

In this paradigm, the model is trained in an enriched feature space – the combined feature spaces of all participants. However, a list of overlapped users should be created to arrange such training. Companies do not want to expose other users. To make the list of intersections, encryption-based entity alignment methods are used [29,30]. The method adds to the schema one more participant acting as an arbiter. All other participants should trust it. Without loss of generality, suppose we have two participants in the

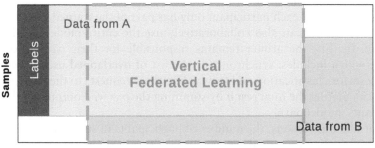

Figure 1.3 Vertical (feature-partitioned) federated learning. (Based on [80].)

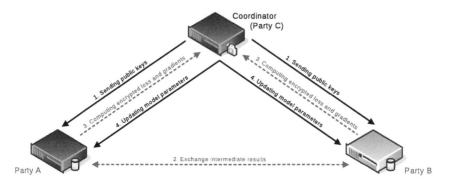

Figure 1.4 Vertical federated learning with a coordinator. (Based on [34].)

training: A, which keeps its part of the features and has all the labels; and B, which only has its part of the features. The arbiter (or coordinator) is denoted by C (Figure 1.4).

Once the list of intersections has been created, the next step is the training itself. Generally, parts of the model belonging to different participants do not have to be equal. The input of each participant's model part is its feature set. The arbiter composes the next batch, each participant runs a learning step, and after that, the sum of the output values of all parts of the model is used to calculate the loss [28]. The additively homomorphic encryption [31,32] with the coordinator's public key is used to ensure robust security while exchanging the outputs. It allows the participant with labels to calculate the loss in an encrypted form and transfer this value to the rest of the participants; after that, all the participants can figure out the encrypted values of gradients for their parts of the model. Next, these gradients (with salt so that the arbiter cannot find their actual values) and losses are received and decrypted by the arbiter. The decrypted gradients are sent to their owners, and the loss value is used to monitor the training progress. The participants then update the parameters of their models and start the next training step. At the validation step, the arbiter uses the loss to check the condition of the model's convergence.

After training, each participant only has part of the trained model. Thus, the participants can also collaboratively use the entire model in inference mode. In this, the arbiter remains responsible for their work coordination, which includes: synchronizing the list of overlapped users; accepting requests for classification and forwarding these requests to the participants; and calculating the final result by summing the received outputs of the participants' model parts.

For obvious reasons, the number of participants in such settings is small. Most of the literature analyzes the cases of two participants and an arbiter. However, for example, [33] proposes a framework for multi-class VFL

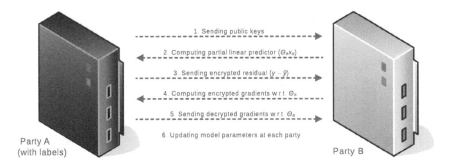

Figure 1.5 Vertical federated learning without a coordinator. (Based on [34].)

involving multiple parties (Multi-participant Multi-class Vertical Federated Learning (MMVFL) framework).

Finding an arbiter that all project participants can trust is often not easy. In [34], a proposed solution eliminates the third-party coordinator from the scheme (Figure 1.5). This approach leads to a reduction in both organizational and technical system complexity.

The article considers the case of two data holders, but the solution is quite easily generalized to the case of N participants.

1.1.13 Federated transfer learning

In addition to the horizontal and vertical learning mechanisms discussed above, a more general case is possible when organizations' data have little overlap in terms of both customer lists and feature sets. For that case, [35] proposes an approach called *federated transfer learning* (FTL) (Figure 1.6).

Let us say (without loss of generality) that organization A has a tagged dataset and a self-learned model that has been trained on that dataset. At the same time, organization B has its own – unlabeled – dataset, with a feature set different from A but some intersections with A regarding the

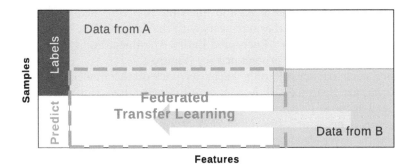

Figure 1.6 Federated transfer learning. (Based on [81].)

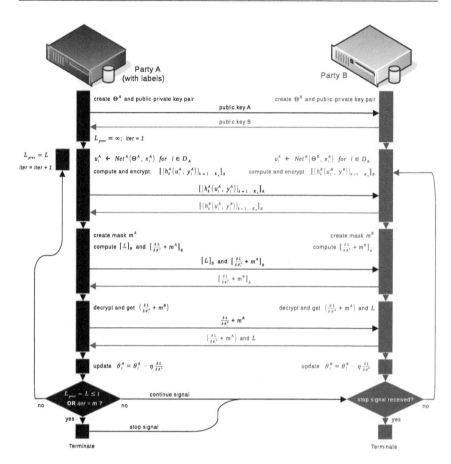

Figure 1.7 Homomorphic encryption-based FTL algorithm workflow. (Based on [35].)

customers. In that case, one can add to the model of A an additional part on the side of B, which will translate features of B into the internal representation of the model of A, and fine-tune the combined model on the part of the dataset of B that intersects with the dataset of A. As in the case of VFL, both algorithms, Encrypted Entity Alignment and Homomorphic Encryption, are used to protect privacy (Figure 1.7).

Since each participant only keeps its part of the model, cooperation between participants is still necessary at inference time, similar to VFL.

The workflow pictured in Figure 1.7 contains only actors A and B. However, if necessary, the interaction protocol can be easily changed to add a third party – an arbiter [36]. For that use case, the workflow is exactly as shown in Figure 1.4.

1.2 SPLIT LEARNING AND SPLIT
FEDERATED LEARNING

The main reason for FL's emergence was the need to protect data privacy. However, the most commonly used horizontal FL approach has its drawbacks. These are, in particular, the requirements for the transmission medium. As state-of-the-art Deep Learning models reach enormous sizes, transferring such models to and from the clients at the beginning and end of each training round imposes high minimum requirements on the speed and reliability of the clients' Internet connection. Compression algorithms can slightly reduce these requirements, but this does not generally affect the situation.

Some learning algorithms reduce the frequency of data transfers by increasing the number of epochs running on the client during one round. However, the statistical heterogeneity of the clients' data degrades, sometimes significantly both the convergence process and the resulting model performance. For example, in the classical algorithm FedAvg [11], which was first proposed to run several training epochs on the client in one round, model convergence is not guaranteed if the number of epochs is set too large. Algorithms like FedProx [12] or SCAFFOLD [13] try to fix this problem with some corrections, which work to a certain extent. However, the models continue to grow in size, so it is impossible to reduce the significance of the communication factor.

In addition to high requirements for the capabilities of transmission media, models of large size impose strong requirements on clients' storage and processing capacity. If we consider the cross-device setting, state-of-the-art models often simply do not fit in the memory of edge devices. Moreover, training such models requires a lot of computational power. As a result, we can either select only the most powerful devices as clients, cutting off all the others, or reduce the complexity of the trained models and perform their quantization, distillation, and so on.

When organizing a cross-silo setting, we can face similar difficulties. In this setting, clients have much larger local datasets, so participants may need special hardware to handle machine learning tasks for training to be completed within a reasonable time frame, for example, standalone servers with GPUs. These additional costs may not fit into the budgets of some potential participants.

Another drawback of horizontal FL, inherent in this technology by design, is that although we keep user data privacy, the model itself – its structure and parameters – is public information transferred during training to clients' devices. Sometimes, the training organizer does not want competitors to have access to this information. While in private FL systems (the cross-silo setting), all training participants may be motivated to keep the model secret, offering a simple solution for public FL systems (the cross-device setting) is challenging.

1.2.1 Vanilla split learning

To address some of those challenges, [37] proposed *splitNN:* an approach under distributed collaborative machine learning paradigm for assembling deep neural networks in a cross-silo setting. The mechanism to train such models was called *split learning* (SL) [38].

SplitNN is a deep neural network that is vertically split into a few sections. Each section includes several layers of the original DNN and is either on the client- or server-side.

The configuration in Figure 1.8 is called *simple vanilla*, and during the training it works as follows. The client forward propagates the next batch through its model section until the *cut* (or *split*) *layer.* Then, the output from cut layer activation, known as *smashed data* [39], is transferred to the server and forward propagates through its section. After that, the server calculates a loss, backpropagates through the layers of its section, and finally gets the gradients for client-cut layer activation. This data is transferred back to the client, and it backpropagates the data through its section. In the inference time, only the forward propagation step is run.

Obviously, in the case when there is only one client (e.g., the client has enough data, but it rents resources for training, for example, in the cloud), the quality of training of such a model does not differ from the model without split (which is entirely located on the client). For the case of several clients, [37] proposes a serial learning algorithm (*alternate client training* [40]), where one client is trained and the rest wait for their turn. In this case, all clients use the same configuration for their model sections, and before

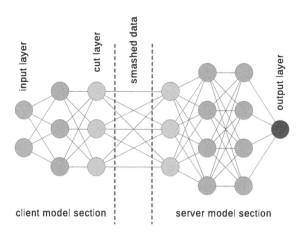

Figure 1.8 Simple vanilla split learning.

starting the training epoch, the next client first receives and loads into its model section the parameters passed by the client that previously completed its turn. Two modalities of the splitNN algorithm are proposed: *centralized mode* and *peer-to-peer mode*. In the first case, the model parameters are transferred through the server, while in peer-to-peer mode, the parameters are sent directly between clients, and the server only points to which clients have participated.

References [40–42] show that because of non-IID data sources, the serial training method used in SL can lead to what is technically termed as "catastrophic forgetting," where the trained model highly favors the client data it recently used for training. Experimental results in [43] show that increasing the number of clients negatively affects performance.

Reference [44] proposes another training method called *alternate minibatch training* to mitigate this effect. According to this, the transfer of parameters and switching to a new client do not occur at the end of the entire training epoch of the previous client, but after this client has finished processing just one batch. The authors show an increase in performance, but this method takes more time to train and requires a high-speed network with large bandwidth since the amount of data transferred with this approach is enormous. Here we note that the overall data transfer in SL for any approach is usually higher than in HFL since the cross-silo setting assumes that each client has a relatively large dataset. Furthermore, while model parameters in HFL, albeit of a large size, are passed from/to the server only at the beginning and end of each training round, in SL, smashed data is transferred with each batch processing, so the total size of transmitted data directly depends on the sizes of client datasets.

If the project conditions allow this possibility, clients can attach the server section of the model to their own after completing the training and use the joint model locally and independently of other participants.

1.2.2 Configurations of split learning

The authors [37,38] propose three basic SL configurations (Figure 1.9).

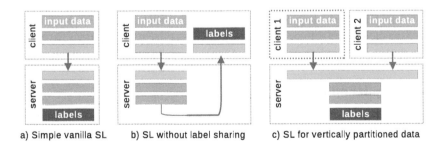

a) Simple vanilla SL   b) SL without label sharing   c) SL for vertically partitioned data

Figure 1.9 Basic split learning configurations. (Based on [38].)

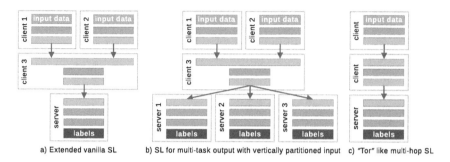

a) Extended vanilla SL   b) SL for multi-task output with vertically partitioned input   c) "Tor" like multi-hop SL

Figure 1.10 Extended split learning configurations. (Based on [38].)

Simple vanilla is an elementary configuration and was discussed above.

SL without label-sharing or *U-shaped configuration* is used when model labels contain sensitive information. As in all other configurations, the server does the majority of work, but the final layers of the model are again located on the client, which calculates the loss and starts backpropagation without sharing the labels.

SL for vertically partitioned data allows multiple clients holding data from different feature spaces to train the model collaboratively without sharing data. Before propagating through the server section, smashed data from the cut layers of both client sections are concatenated. All clients' data must be synchronized and be characteristics of the same entities. (We have already addressed this issue when discussing Vertical FL.)

This configuration list is not exhausted. For different purposes, one can offer more suitable options. For example, [38] provides three additional configurations (Figure 1.10).

Extended vanilla SL: The result of concatenated outputs is further processed by another client before being sent to the server. If all the clients are different departments of the same organization, the approach adds more privacy to the organization's data.

SL for multi-task output with vertically partitioned input: The extended vanilla where the concatenating client sends smashed data over to multiple servers, each of which trains its model section to solve a separate learning task.

"Tor" like multi-hop SL: The extended vanilla configuration in which the smashed data from concatenating clients consequently passes through a few more clients before landing on the server, which completes the training.

1.2.3 SplitFed learning

As we can see, SL solves the lack of storage and computational resources. It could also significantly reduce the requirements for the communication capacity of the clients with small datasets (the cross-device case) since, in such a scenario, the gradients from one or two layers of the model that need to be transferred over the network take up much less space than the complete set of model parameters. Another advantage is the privacy of the server section of the model, which is kept on the server during the entire training and is not shared anywhere.

However, it is impossible to use SL in a cross-device setting because this mechanism assumes that clients are trained sequentially. Since datasets on edge devices are small, the overall latency of switching between a considerable number of clients becomes unacceptably high. In addition, as we mentioned above, there is a high probability that some devices may drop out during the training round, resulting in the wall-clock time of learning being stretched even more.

To eliminate this hindrance, the *Split Federated Learning* (SFL) or *SplitFed* mechanism was proposed [45], combining the advantages of both FL and SL approaches.

For SplitFed, the authors added one more entity called *fed server* to SL. At the start of training, this server provides all clients with client sections of the model that are identical in structure and initial state. At the end of each training round, the clients who took part in this round send updates of their sections to the fed server. Afterward, the server aggregates these updates and sends updated parameters to the clients selected to participate in the following training round. Since the initial parameters of all client sections are identical at the beginning of each round, training can be performed in parallel.

With this approach, the amount of data transmitted over the network is significantly lower than with HFL since the client sections are much smaller than the whole model. As a result, the network latency and the wall-clock learning time are significantly lower. At the same time, all the benefits of the SL approach are preserved, namely, the reduction of technical and communication requirements for the devices participating in the training and the protection of the privacy of the structure and parameters of the server section of the model, which are not transferred anywhere. In addition, it should be noted that any of the SL configurations can be used as a basis when organizing a SplitFed setting.

In [45], two options for organizing SFL are proposed: *SplitFedv1* (SFLv1) and *SplitFedv2* (SFLv2). In SFLv1, clients run forward propagation in parallel, send smashed data to the server, and then the server performs forward and backward propagation steps in parallel for each client on a separate copy of its model section and sends the gradients back to respective clients. Clients run a backpropagation step, and the server this time aggregates the new parameters of trained copies of its section. The fed server uses the same algorithm as the main server (the authors mention a weighted average) to aggregate updates of the client sections of the model, so when the round has ended, both client and server parts of the model are synchronized.

SFLv2, in contrast, updates the server model section with forward-backward propagation sequentially concerning the client's smashed data. The client's order for this is chosen randomly. The fed server then aggregates clients' updates (the client's part of the algorithm is no different from the SFLv1).

Experiments in [45] show that models trained with SPLv1 or SPLv2 have a similar quality to SL and HFL. However, the wall-clock time of learning in SPLv1 is significantly less. Additionally, "catastrophic forgetting" because of serial training on non-IID data [40–42], which we have mentioned about the SL approach, also occurs for SPLv2. However, this effect is hardly notable in the cross-device setting: due to the small size of client datasets, the approach is very close to the SL alternate minibatch training method discussed above.

Additionally, [40] proposes one more architecture called SplitFedv3 (SFLv3). With this approach, the server-side network is an averaged version, the same as in SplitFedv1, and the client-side networks are unique for each client. Fed server is absent in this architecture, so we have a variant of the SL approach where "catastrophic forgetting" is avoided due to averaging the server-side network. After learning, each client keeps its own client section, so this approach cannot be extended to the cross-device setting. The authors show that the performance of this method is slightly higher than that of SL and SplitFedv2, but cannot beat HFL. Non-label-sharing configurations (U-shaped) perform slightly worse than label sharing.

The SFLv1 architecture still seems to be the best solution for cross-device settings.

1.3 DECENTRALIZED (PEER-TO-PEER) FEDERATED LEARNING

A central server can sometimes be considered a bottleneck in the standard FL paradigm. A failure of the central server makes the whole setting non-operational. In addition, the organization that controls the central server may try to reconstruct private client data using their updates. Below, we will touch upon the topic of security, but here we note that since clients communicate in centralized FL only with the server, a malicious server jeopardizes all participants' privacy. For this reason, when organizing a cross-silo setting, the question of which side will control the server is often the stumbling block that stagnates the project.

Various decentralized approaches have been proposed to address this challenge. The unifying point between them is the absence of a central server. Usually, in such settings, communications are carried out between neighboring (in the sense of topology) clients of the computational graph, and the topology itself can be pretty complex.

The topic of *decentralized* (or *peer-to-peer*) *FL* is currently in an active research phase. However, although the industry has not yet developed any standards in this area, most of the proposed solutions to date use Blockchain mechanisms (see, for example, [46–49]). In such architectures, smart contracts are responsible for the tasks performed by the server in centralized FL settings.

It is possible to use other methods of decentralization [50,51]; however, the Blockchain technology by design already includes mechanisms that can address such core challenges of FL as incentive and security.

As mentioned above, convincing incentive mechanisms for FL have not yet been created. At the same time, Blockchain is the platform that already has built-in tools for calculating each participant's contributions. This can be used as protection against *free riders* when some participants pretend to train the model, but the updates they download do not improve the model's quality. Also, this instrument can limit the functionality obtained by the participant's model instances following their contribution (although such segregation may incentivize some potential participants not to participate).

An additional incentive mechanism for the cross-device setting is the possibility of using the native cryptocurrency of the Blockchain project as payment for contribution. This can help attract data owners into the project who are not interested in accessing the final model but just want to make a profit. Similarly, non-data owners can be involved in using the computational power of their devices [52] for intermediate calculations in distributed models [53], searching for malicious activities of other participants [54], etc.

The cryptocurrency's real value can be established based on the results of the commercial use of the models learned within the project. As proposed in [55], learning markets with auditability and traceability mechanisms for trading trusted and verified models is also an exciting idea.

In Blockchain, each transaction, before being added to the ledger, must be verified by the majority of participants. This requirement ensures security and data integrity when all participants hold identical ledger copies. Such architecture "brings trust to a trustless environment" [56]. *The system is secure as long as honest nodes collectively control more CPU power than any cooperating group of attacker nodes* [57].

However, as already mentioned, any particular decentralized architecture that would become the de facto standard for organizing peer-to-peer FL solutions has not yet been found. This is an obstacle to the practical use of the concept. Organizations that would like to take part in a cross-silo FL project often lack the experience to choose, from the many architectures and frameworks proposed at the moment, the one that indeed provides the project with the necessary level of privacy and security.

In the cross-device setting, there are no big projects either. Here, the reason we see is a bit different. The organizers of a distributed FL project can choose any architecture, but what kind of project could it be? It must be attractive enough from a financial point of view so that the profit from its implementation can motivate device owners to join. However, if such a project is found, the companies that own popular applications installed on billions of devices would be able to implement it themselves without even asking users for permission. After that, the distributed project would lose its meaning since a large corporation can always offer a lower price and better support to those who want to access the project results.

1.4 SUMMARY

Although now the primary audience of decentralized FL is researchers and enthusiasts, in a cross-silo setting, the technology prospects seem quite good. Equal partnership – without nominating any of the participants in the lead role and transparent rules for evaluating each participant's contribution – may be (and already turns out to be) an effective incentive mechanism. A significant increase in the number of such projects may occur after solutions based on a specific decentralized FL architecture begin to prevail and thus become the de facto standard. In a cross-device setting, the possibility of widespread use of the technology is still questionable.

1.5 NLP THROUGH FL

If necessary, FL can be applied to any NLP task; however, the problems addressed in the cross-device setting are usually of the language modeling type. Almost the only valuable textual information sources on mobile devices are private messages that users exchange in instant messengers or some private social content less often. All other texts are usually put online, and it is easier to train models on them centralized.

In FL, personal messages were used to train the models that: suggest the next typing word or tailing of the currently typed word on the keyboard [21], make emoji suggestions [58], and learn out-of-vocabulary words [59]. It is also possible to create models that offer ready-made user answers to an incoming message. Since suggestion functionality is a part of virtual keyboards, these models are primarily of interest to companies – developers of mobile devices, which, of course, have all the conditions to arrange such FL projects. So, it is unsurprising that Google submitted all the articles mentioned in this paragraph. Any mobile device user may assess the current quality of the suggested models.

Also, when preparing texts of any type (personal messages, texts intended for further publication, etc.), users usually correct typos and style. The history of these corrections may be used to train the model to find poorly worded text constructs and suggest replacements.

All the tasks mentioned above can be addressed using an unsupervised learning approach. However, having access to the device owners' personal information (region of residence, education, place of work, position, age, social circle, etc.), one could use these data as labels to arrange supervised classification. This would allow detecting the essential features of a person by the style of its text. Then, these models could be used, for example, to optimize chatbots by choosing the most convenient communication style for the caller and assessing the veracity of the caller's information about himself. There are many applications for such models.

Mobile devices have a limited storage capacity. If the model is supposed to work standalone on the device, state-of-the-art transformer-based models cannot be used. So far, suggestion models have been developed on RNN or CNN (word- or character-level) architectures. However, for models the use of which is not related to downloading on mobile devices, it is possible to arrange through FL (for example, using the FSL approach) fine-tuning of the state-of-the-art NLP models, which often may be done in a one- or few-shot manner.

In the cross-silo setting, resource constraints are less relevant; therefore, within this setting, it is possible to handle any task for which the organizations or data centers involved have labeled datasets. Since the performance of state-of-the-art models and the model's size mainly depend on the training set size [60], increasing this size through cooperation, even using the FL paradigm, makes sense.

For example, [61] describes experiments with fine-tuning the BERT model on a set of clinical notes. As an extrinsic evaluation task, the authors used NER modeling. This experiment's relevance seems dubious since the authors used the freely available dataset MIMIC-III [62], which was "randomly split into 5 groups to mimic 5 different silos by the patient". Apparently, in this case, the data across all the silos can be considered identically and independently distributed, so it is not surprising that the quality of the model did not differ much from that obtained using centralized learning.

Generally speaking, a relatively small dataset is sufficient to fine-tune an already trained transformer model. Furthermore, if we take instead a large dataset whose distribution across silos is significantly heterogeneous, we can expect a deterioration in the resulting performance. Whatever the case, comparing the quality of self-learned local models with that obtained using FL makes sense.

Actually, in healthcare, FL finds the most widespread use. The reason is extreme restrictions on data transfer, even between departments of the same hospital. Intuitively, statistical heterogeneity within this domain may not be as pronounced as among the open-domain datasets. Due to this, NLP tasks in this field are addressed through FL quite efficiently [63–65].

Reference [66] describes the application of FL to address the NER task of tagging specific terms in medical texts. The CNN–biLSTM–CRF model accepts as input a concatenation of three types of embeddings: Glove word-level embeddings, character-level embeddings produced by the CNN extra layer, and contextual embeddings from a pre-trained ELMo language model. Statistical heterogeneity is achieved by using different datasets in different silos. Since these datasets are incompatible with the notation and compilation principles, the authors make the biLSTM–CRF model head private for each silo. During training, only the lower layers are shared. The authors show that the resulting model significantly outperforms those trained on only one of the datasets. Therefore, the authors used FL to train a contextual embedding model of their architecture, and the remaining part of the model served for extrinsic evaluation using multiple NER tasks. As a result, training such a model from scratch (with frozen ELMo layers) delivers good performance even if the dataset is non-IID.

However, the task of learning from scratch the state-of-the-art contextual embedding model (BERT, etc.) is very tough. This requires too much computational power (preferably a TPU) and a giant dataset. Even with a small portion of such a dataset, any potential project participant could take an already pre-trained embedding model for either an open domain or its domain of interest, fine-tune this model with only its dataset, and probably achieve better results than through a complicated FL setting.

Here we would like to highlight the recent release of the FedNLP framework [67], a research platform for studying the possibilities of using FL in NLP. It has a comprehensive list of task formulations for NLP applications and a unified interface between FL methods provided by *FedML* research and benchmarking framework [68] and primarily transformer-based language models provided by the *Hugging Face*'s *transformers* library [69]. It also contains evaluation protocols with comprehensive partitioning strategies for simulating non-IID client distributions. Evaluating current state-of-the-art NLP models, the authors register that federated fine-tuning still has a large accuracy gap in the non-IID datasets compared to centralized fine-tuning. Moreover, since fine-tuning of pre-trained models is now part of almost any NLP pipeline, the widespread use of FL in NLP is hard to imagine without bridging this gap.

1.6 SECURITY AND PRIVACY PROTECTION

So far, we have ignored privacy protection issues, although the very need to protect data privacy has become the main reason for the emergence of the entire FL paradigm. The security threats typical for FL can be categorized according to the **CIA Triad** security model: **Confidentiality, Integrity,** and **Availability.** In particular, the *confidentiality attack* aims to reveal any sensitive information; the *integrity attack* tries to put into the resulting model a backdoor that implements an incorrect, but desirable for the attacker, reaction of the model to the given combinations of input parameters; and the *availability attack* just sets a goal to diminish the resulting model's performance. Integrity and availability attacks together are also known as *poisoning attacks*.

Communication security methods solve the issues when the attacker is outside the FL setting, and we will not enter into that. The situation becomes much more complicated if an attacker either manages to get inside the FL setting or is initially a part of it. The FL security model deals with situations of s*emi-honest (honest-but-curious)* and *malicious* client(s) or server(s).

The semi-honest participant operates in full accordance with the FL protocol, i.e., it correctly serves its role in the setting for training the model. Additionally, it analyzes intermediate data received from other participants to reconstruct the information for the sake of hiding, for which all this FL setting was established. In other words, that participant performs the confidentiality attack. Because this is a passive attack, it is impossible to detect the presence of such an adversary in the setting. Therefore, in the FL paradigm, all participants, both clients and the server, are a priori considered semi-honest, and in the learning process, mutual data transfer between the participants should be minimized. For example, in HFL, the client most often communicates only with the server and does not see updates sent by its neighbors in the setting. The server, however, sees all updates.

The malicious participant seeks to influence the final functionality of the model. The goal can be to degrade the resulting performance or insert a backdoor, e.g., for identifying any person wearing a bow tie as a full-access employee. Even in cases where an attacker violates the FL protocol and actively interferes with the training process, its disclosure is complicated and not always feasible. For example, we can detect a client sending updated parameters that are too different from the current parameters of the model. However, nothing prevents an attacker from adding an instrumental term to the objective function that penalizes these differences.

The case of the malicious server can theoretically be recognized. For example, if we connect to another server that is nt affiliated with the first one, we can compare the results of their work. However, in this case, we can get one more honest-but-curious server in the system and thus double the probability of data leakage. By now, quite a few methods of reverse engineering the data based on the gradients of the learning model have been demonstrated. For example, the method proposed in [70] allows obtaining the training inputs and the labels in just a few iterations. The proposed algorithm can recover pixel-wise accurate original images and token-wise matching original texts. Some discrepancies in the algorithm were corrected in [71].

As a by-product, some methods used in model training, such as gradient clipping or dropout, complicate reverse engineering. However, such protection is insufficient. This can only require more time for the attacker to obtain the desired information, not prevent it. The server can continue learning as long as it needs to, so regularization methods do not solve the problem. It is possible to completely protect the data using homomorphic encryption algorithms [72] or secure multiparty computation [73,74]; however, these algorithms require a massive amount of additional resources. For example, logistic regression on 1 MB of data requires 10 GB of memory, and massive parallelization is necessary to reduce computation latency to a value acceptable for practical tasks [75].

A more usable method is based on differential privacy [76]. Before sending updates, the client adds noise to them. More noise variance means higher protection. Nevertheless, this also means lower model performance. There is always a trade-off between these aspects. The amount of noise that can deliver absolute privacy protection does not allow the model to converge.

The integrity and availability attacks, as we describe them, are not peculiar FL problems. In the compromised system, a model made with centralized training may be poisoned similarly. In FL, the situation is even less complicated. Of course, if the server is compromised, there is no protection. However, if an honest server has received a poisoned update from some client, the malicious effect can be mitigated to some degree after aggregating all client updates. Therefore, the more clients in the setting, the less negative the impact of one adversary on the entire system. Nevertheless, if the attacker succeeds in seizing several clients at once, their coordinated impact can prevail over passive tools used by the server (Byzantine attack [77]).

For active response, the server itself can apply the differential privacy methods. Namely, to add noise during the aggregation of client updates. Such methods make a malicious task much more complex for the attacker. However, as mentioned above, it negatively affects the model's performance. Furthermore, if clients already use DP methods to protect against the confidentiality attack, further use of these methods on the server can make the resulting model's performance too low.

In the cross-device setting, when the server manages the entire process and clients may not even know that they are involved in some training, the problem of the internal adversary may be ignored. Malicious clients, if they appear, dissolve into a vast number of honest ones. Furthermore, an honest-but-curious server that also controls the software on the client side has no reason to implant protection in this software against the server itself. No one doubts that the server does not reconstruct clients' data from their updates. Indeed, if it wants the client's raw data, it can download it silently, right?

In addition to those discussed, there are also inference-time attacks aimed at obtaining unforeseen information from the model, finding ways to disrupt its behavior, reconstructing the internal structure of the model based on only its inputs and outputs, etc. However, these attacks do not differ from similar attacks on models obtained by centralized learning, and we do not consider them here.

1.7 FL PLATFORMS AND DATASETS

Many software has been designed to simplify the arrangement of FL settings. Some were created and continue to be developed primarily as tools to accelerate research. They allow creating and testing different use cases that are rarely found in natural production systems. Others are specifically aimed at ensuring the smooth operation of production-level systems. Those interested can be redirected to Appendix A of [2], which contains an almost exhaustive FL software list.

Additionally, we would like to draw attention to the *OpenFL* framework recently presented by Intel [78]. It is an open-source project that supports *Tensorflow* and *PyTorch* pipelines in a cross-silo setting and strongly emphasizes security. One of the key features is an effort to keep the model structure secret from clients using trusted execution environments. This approach significantly increases the cost of arranging poisoning attacks for a malicious client. Suppose the planned setting has a server that potential participants trust, then using TEE on clients can be an additional incentive for them to join the project. Of course, if there is no such server, then the lack of ability to control the learning process is a negative incentive for potential clients. However, it is worth noting that TEE is organized using Intel Software Guard Extensions (Intel SGX), provided by the latest Intel Scalable Xeon processors, which imposes severe hardware requirements on the participants in such an FL setting (perhaps this should have been said at the beginning of the paragraph).

As to FL datasets, many benchmarking datasets for different tasks, including some NLP tasks, have recently been presented in the *FedScale* collection [79]. These are already known datasets, but now they have all been collected into one library and supplemented by *FedScale Automated Runtime* to simplify and standardize the FL experimental setup and model evaluation process.

1.8 CONCLUSION

FL, in many aspects, is a forced technology brought to life by the presence of various kinds of regulations, the by-product of which is the impossibility of arranging some ML projects in a "traditional" way. The performance of FL models can often be considered acceptable, but compared to centralized learning, this is a step backward. In addition, progress does not stop, and more advanced models appear, but if, with centralized training, one can simply train them on the old dataset, then access to the old dataset is most likely impossible in the FL paradigm. Again, we must organize a collaborative learning setting, address administrative issues, etc.

However, we would like to note that progress also continues in the area of FL. New algorithms for aggregation and privacy protection appear.

The main FL challenge – statistical heterogeneity – is not very pronounced in some industries, so now FL already makes it possible to solve quite an extensive range of problems with reasonably good quality. Moreover, as FL technology improves, the quality of its applications will only grow in the future.

REFERENCES

[1] Sweeney, L. (2000). *Simple Demographics Often Identify People Uniquely.* Carnegie Mellon University. Journal contribution.

[2] Kairouz, P., McMahan, H.B., Avent, B., Bellet, A., Bennis, M., Bhagoji, A., Bonawitz, K., Charles, Z.B., Cormode, G., Cummings, R., D'Oliveira, R.G., Rouayheb, S., Evans, D., Gardner, J., Garrett, Z., Gascón, A., Ghazi, B., Gibbons, P.B., Gruteser, M., Harchaoui, Z., He, C., He, L., Huo, Z., Hutchinson, B., Hsu, J., Jaggi, M., Javidi, T., Joshi, G., Khodak, M., Konecný, J., Korolova, A., Koushanfar, F., Koyejo, O., Lepoint, T., Liu, Y., Mittal, P., Mohri, M., Nock, R., Özgür, A., Pagh, R., Raykova, M., Qi, H., Ramage, D., Raskar, R., Song, D., Song, W., Stich, S.U., Sun, Z., Suresh, A.T., Tramèr, F., Vepakomma, P., Wang, J., Xiong, L., Xu, Z., Yang, Q., Yu, F., Yu, H., & Zhao, S. (2021). Advances and Open Problems in Federated Learning. *Foundation and Trends in Machine Learning, 14,* 1–210.

[3] Dwork, C. (2006). Differential Privacy. *2006 33rd International Colloquium on Automata, Languages and Programming,* 1–12.

[4] Yao, A. (1982). *Protocols for Secure Computations.* FOCS.

[5] Yao, A. (1986). How to Generate and Exchange Secrets. *27th Annual Symposium on Foundations of Computer Science (SFCS 1986),* 162–167.

[6] Goldreich, O., Micali, S., & Wigderson, A. (1987). How to Play ANY Mental Game. *Proceedings of the Nineteenth Annual ACM Symposium on Theory of Computing.*

[7] Pathak, M.A., Rane, S., & Raj, B. (2010). *Multiparty Differential Privacy via Aggregation of Locally Trained Classifiers.* NIPS.

[8] Shokri, R., & Shmatikov, V. (2015). Privacy-Preserving Deep Learning. *2015 53rd Annual Allerton Conference on Communication, Control,* and Computing *(Allerton),* 909–910.

[9] Konecný, J., McMahan, H.B., & Ramage, D. (2015). Federated Optimization: Distributed Optimization Beyond the Datacenter. ArXiv, abs/1511.03575.

[10] McMahan, H.B., Moore, E., Ramage, D., Hampson, S., & Arcas, B.A. (2017). *Communication-Efficient Learning of Deep Networks from Decentralized Data.* AISTATS.

[11] Nilsson, A., Smith, S., Ulm, G., Gustavsson, E., & Jirstrand, M. (2018). A Performance Evaluation of Federated Learning Algorithms. DIDL'18. *Proceedings of the Second Workshop on Distributed Infrastructures for Deep Learning,* December 2018, 1–8. DOI 10.1145/3286490.3286559

[12] Sahu, A.K., Li, T., Sanjabi, M., Zaheer, M., Talwalkar, A.S., & Smith, V. (2020). Federated Optimization in Heterogeneous Networks. ArXiv, 1812.06127.

[13] Karimireddy, S.P., Kale, S., Mohri, M., Reddi, S.J., Stich, S.U., & Suresh, A.T. (2020). *SCAFFOLD: Stochastic Controlled Averaging for Federated Learning.* In: Daume, H. & Singh, A. (eds.), *37th International Conference on Machine Learning, ICML 2020,* ICML 2020; Vol. PartF168147-7, 5088–5099. International Machine Learning Society (IMLS).

[14] Reddi, S.J., Charles, Z.B., Zaheer, M., Garrett, Z., Rush, K., Konecný, J., Kumar, S., & McMahan, H.B. (2021). Adaptive Federated Optimization. ArXiv, abs/2003.00295.

[15] Zhang, J., Karimireddy, S.P., Veit, A., Kim, S., Reddi, S.J., Kumar, S., & Sra, S. (2020). Why are Adaptive Methods Good for Attention Models? ArXiv, 1912.03194.

[16] Defazio, A., & Bottou, L. (2019). On the Ineffectiveness of Variance Reduced Optimization for Deep Learning. NeurIPS.

[17] Ji, S., Pan, S., Long, G., Li, X., Jiang, J., & Huang, Z. (2019). Learning Private Neural Language Modeling with Attentive Aggregation. *2019 International Joint Conference on Neural Networks (IJCNN)*, 1–8.

[18] Wang, J., Liu, Q., Liang, H., Joshi, G., & Poor, H. (2020). Tackling the Objective Inconsistency Problem in Heterogeneous Federated Optimization. ArXiv, abs/2007.07481.

[19] Zhang, X., Hong, M., Dhople, S., Yin, W., & Liu, Y. (2020). FedPD: A Federated Learning Framework with Optimal Rates and Adaptivity to Non-IID Data. ArXiv, abs/2005.11418.

[20] Acar, D.A., Zhao, Y., Navarro, R.M., Mattina, M., Whatmough, P., & Saligrama, V. (2021). *Federated Learning Based on Dynamic Regularization*. ICLR.

[21] Hard, A., Rao, K., Mathews, R., Beaufays, F., Augenstein, S., Eichner, H., Kiddon, C., & Ramage, D. (2018). Federated Learning for Mobile Keyboard Prediction. ArXiv, abs/1811.03604.

[22] Wu, Q., He, K., & Chen, X. (2020). Personalized Federated Learning for Intelligent IoT Applications: A Cloud-Edge Based Framework. *IEEE Open Journal of the Computer Society*, 1, 35–44.

[23] Mansour, Y., Mohri, M., Ro, J., & Suresh, A.T. (2020). Three Approaches for Personalization with Applications to Federated Learning. ArXiv, abs/2002.10619.

[24] Kulkarni, V., Kulkarni, M., & Pant, A. (2020). Survey of Personalization Techniques for Federated Learning. *2020 Fourth World Conference on Smart Trends in Systems, Security and Sustainability (WorldS4)*, 794–797.

[25] Liu, Y., & Wei, J. (2020). Incentives for Federated Learning: A Hypothesis Elicitation Approach. ArXiv, abs/2007.10596.

[26] Zhan, Y., Zhang, J., Hong, Z., Wu, L., Li, P., & Guo, S. (2021). A Survey of Incentive Mechanism Design for Federated Learning. *IEEE Transactions on Emerging Topics in Computing*, 10, 1035–1044.

[27] Zeng, R., Zeng, C., Wang, X., Li, B., & Chu, X. (2021). A Comprehensive Survey of Incentive Mechanism for Federated Learning. ArXiv, abs/2106.15406.

[28] Yang, Q., Liu, Y., Chen, T., & Tong, Y. (2019). Federated Machine Learning. *ACM Transactions on Intelligent Systems and Technology (TIST)*, 10, 1–19.

[29] Christen, P. (2012). *Data Matching: Concepts and Techniques for Record Linkage, Entity Resolution, and Duplicate Detection*. Peter Christen.

[30] Schnell, R., Bachteler, T., & Reiher, J. (2011). A Novel Error-Tolerant Anonymous Linking Code. November 16, 2011. Available at SSRN: https://ssrn.com/abstract=3549247 or http://dx.doi.org/10.2139/ssrn.3549247.

[31] Rivest, R.L., & Dertouzos, M. (1978). On Data Banks and Privacy Homomorphisms. In: DeMillo, R.A. (ed.), *Foundations of Secure Computation*, Academic Press, 169–179.

[32] Paillier, P. (1999). Public-Key Cryptosystems Based on Composite Degree Residuosity Classes. In: Stern, J. (eds) *Advances in Cryptology – EUROCRYPT '99. EUROCRYPT 1999. Lecture Notes in Computer Science*, vol 1592. Springer. https://doi.org/10.1007/3-540-48910-X_16.

[33] Feng, S., & Yu, H. (2020). Multi-Participant Multi-Class Vertical Federated Learning. ArXiv, abs/2001.11154.

[34] Yang, S., Ren, B., Zhou, X., & Liu, L. (2019). Parallel Distributed Logistic Regression for Vertical Federated Learning without Third-Party Coordinator. ArXiv, abs/1911.09824.

[35] Liu, Y., Kang, Y., Xing, C., Chen, T., & Yang, Q. (2020). A Secure Federated Transfer Learning Framework. *IEEE Intelligent Systems, 35*, 70–82.

[36] Aledhari, M., Razzak, R., Parizi, R., & Saeed, F. (2020). Federated Learning: A Survey on Enabling Technologies, Protocols, and Applications. *IEEE Access, 8*, 140699–140725.

[37] Gupta, O., & Raskar, R. (2018). Distributed Learning of Deep Neural Network over Multiple Agents. *Journal of Network and Computer Applications, 116*, 1–8.

[38] Vepakomma, P., Gupta, O., Swedish, T., & Raskar, R. (2018). Split Learning for Health: Distributed Deep Learning Without Sharing Raw Patient Data. ArXiv, abs/1812.00564.

[39] Vepakomma, P., Gupta, O., Dubey, A., & Raskar, R. (2019). *Reducing Leakage in Distributed Deep Learning for Sensitive Health Data*. Presented at the ICLR AI for social good workshop 2019. https://aiforsocialgood. github.io/iclr2019/accepted/track1/pdfs/29_aisg_iclr2019.pdf

[40] Gawali, M., Arvind, C., Suryavanshi, S., Madaan, H., Gaikwad, A., Prakash, K., Kulkarni, V., & Pant, A. (2021). Comparison of Privacy-Preserving Distributed Deep Learning Methods in Healthcare. ArXiv, abs/2012.12591.

[41] Sheller, M.J., Reina, G.A., Edwards, B., Martin, J., & Bakas, S. (2018). Multi-Institutional Deep Learning Modeling Without Sharing Patient Data: A Feasibility Study on Brain Tumor Segmentation. *Brainlesion, 11383*, 92–104.

[42] Sheller, M.J., Edwards, B., Reina, G.A., Martin, J., Pati, S., Kotrotsou, A., Milchenko, M., Xu, W., Marcus, D., Colen, R., & Bakas, S. (2020). Federated Learning in Medicine: Facilitating Multi-Institutional Collaborations without Sharing Patient Data. *Scientific Reports, 10*, 12598.

[43] Madaan, H., Gawali, M., Kulkarni, V., & Pant, A. (2021). Vulnerability Due to Training Order in Split Learning. ArXiv, abs/2103.14291.

[44] Roy, A.G., Siddiqui, S., Pölsterl, S., Navab, N., & Wachinger, C. (2019). BrainTorrent: A Peer-to-Peer Environment for Decentralized Federated Learning. ArXiv, abs/1905.06731.

[45] Thapa, C., Arachchige, P.C., & Çamtepe, S. (2020). SplitFed: When Federated Learning Meets Split Learning. ArXiv, abs/2004.12088.

[46] Kim, H., Park, J., Bennis, M., & Kim, S. (2020). Blockchain On-Device Federated Learning. *IEEE Communications Letters, 24*, 1279–1283.

[47] Shayan, M., Fung, C., Yoon, C.J., & Beschastnikh, I. (2018). Biscotti: A Ledger for Private and Secure Peer-to-Peer Machine Learning. ArXiv, abs/1811.09904.

[48] Weng, J., Weng, J., Li, M., Zhang, Y., & Luo, W. (2021). DeepChain: Auditable and Privacy-Preserving Deep Learning with Blockchain-Based Incentive. *IEEE Transactions on Dependable and Secure Computing, 18*, 2438–2455.

[49] Schmid, R., Pfitzner, B., Beilharz, J., Arnrich, B., & Polze, A. (2020). Tangle Ledger for Decentralized Learning. *2020 IEEE International Parallel and Distributed Processing Symposium Workshops (IPDPSW)*, 852–859.

[50] Domingo-Ferrer, J., Blanco-Justicia, A., Sánchez, D., & Jebreel, N. (2020). Co-Utile Peer-to-Peer Decentralized Computing. *2020 20th IEEE/ACM International Symposium on Cluster, Cloud and Internet Computing (CCGRID)*, 31–40.

[51] Liu, Y., Sun, S., Ai, Z., Zhang, S., Liu, Z., & Yu, H. (2020). FedCoin: A Peer-to-Peer Payment System for Federated Learning. ArXiv, abs/2002.11711.

[52] Hu, C., Jiang, J., & Wang, Z. (2019). Decentralized Federated Learning: A Segmented Gossip Approach. ArXiv, abs/1908.07782.

[53] Bao, X., Su, C., Xiong, Y., Huang, W., & Hu, Y. (2019). FLChain: A Blockchain for Auditable Federated Learning with Trust and Incentive. *2019 5th International Conference on Big Data Computing and Communications (BIGCOM)*, 151–159.

[54] Ouyang, L., Yuan, Y., & Wang, F. (2020). Learning Markets: An AI Collaboration Framework Based on Blockchain and Smart Contracts. *IEEE Internet of Things Journal*, PP(99), 1–1. DOI:10.1109/JIOT.2020.3032706

[55] Rehman, M.H., & Gaber, M.M. (2021). Federated Learning Systems: Towards Next Generation AI. *Studies in Computational Intelligence*, Vol 965. Springer Verlag.

[56] Nakamoto, S. (2009). Bitcoin: A Peer-to-Peer Electronic Cash System. https://bitcoin.org/bitcoin.pdf

[57] Ramaswamy, S.I., Mathews, R., Rao, K., & Beaufays, F. (2019). Federated Learning for Emoji Prediction in a Mobile Keyboard. ArXiv, abs/1906.04329.

[58] Chen, M., Mathews, R., Ouyang, T.Y., & Beaufays, F. (2019). Federated Learning Of Out-Of-Vocabulary Words. ArXiv, abs/1903.10635.

[59] Raffel, C., Shazeer, N.M., Roberts, A., Lee, K., Narang, S., Matena, M., Zhou, Y., Li, W., & Liu, P.J. (2020). Exploring the Limits of Transfer Learning with a Unified Text-to-Text Transformer. ArXiv, abs/1910.10683.

[60] Liu, D., & Miller, T. (2020). Federated Pretraining and Fine-Tuning of BERT Using Clinical Notes from Multiple Silos. ArXiv, abs/2002.08562.

[61] Johnson, A.E., Pollard, T., Shen, L., Lehman, L.H., Feng, M., Ghassemi, M., Moody, B., Szolovits, P., Celi, L., & Mark, R. (2016). MIMIC-III, A Freely Accessible Critical Care Database. *Scientific Data* 3, 160035. https://doi.org/10.1038/sdata.2016.35.

[62] Brisimi, T.S., Chen, R., Mela, T., Olshevsky, A., Paschalidis, I., & Shi, W. (2018). Federated Learning of Predictive Models from Federated Electronic Health Records. *International Journal of Medical Informatics*, 112, 59–67.

[63] Huang, L., & Liu, D. (2019). Patient Clustering Improves Efficiency of Federated Machine Learning to Predict Mortality and Hospital Stay Time Using Distributed Electronic Medical Records. *Journal of Biomedical Informatics*, 99(9), 103291. DOI:10.1016/j.jbi.2019.103291

[64] Cui, J., & Liu, D. (2020). Federated Machine Learning with Anonymous Random Hybridization (FeARH) on Medical Records. ArXiv, abs/2001.09751.

[65] Ge, S., Wu, F., Wu, C., Qi, T., Huang, Y., & Xie, X. (2020). FedNER: Privacy-Preserving Medical Named Entity Recognition with Federated Learning. ArXiv, abs/2003.09288.

[66] Lin, B.Y., He, C., Zeng, Z., Wang, H., Huang, Y., Soltanolkotabi, M., Ren, X., & Avestimehr, S. (2021). FedNLP: A Research Platform for Federated Learning in Natural Language Processing. ArXiv, abs/2104.08815.

[67] He, C., Li, S., So, J., Zhang, M., Wang, H., Wang, X., Vepakomma, P., Singh, A., Qiu, H., Shen, L., Zhao, P., Kang, Y., Liu, Y., Raskar, R., Yang, Q., Annavaram, M., & Avestimehr, S. (2020). FedML: A Research Library and Benchmark for Federated Machine Learning. ArXiv, abs/2007.13518.

[68] Wolf, T., Debut, L., Sanh, V., Chaumond, J., Delangue, C., Moi, A., Cistac, P., Rault, T., Louf, R., Funtowicz, M., & Brew, J. (2019). HuggingFace's Transformers: State-of-the-art Natural Language Processing. ArXiv, abs/1910.03771.

[69] Zhu, L., Liu, Z., & Han, S. (2019). *Deep Leakage from Gradients*. NeurIPS.

[70] Zhao, B., Mopuri, K.R., & Bilen, H. (2020). iDLG: Improved Deep Leakage from Gradients. ArXiv, abs/2001.02610.

[71] Phong, L.T., Aono, Y., Hayashi, T., Wang, L., & Moriai, S. (2018). Privacy-Preserving Deep Learning via Additively Homomorphic Encryption. *IEEE Transactions on Information Forensics and Security, 13*, 1333–1345.

[72] Rouhani, B., Riazi, M., & Koushanfar, F. (2018). DeepSecure: Scalable Provably-Secure Deep Learning. *2018 55th ACM/ESDA/IEEE Design Automation Conference (DAC)*, 1–6.

[73] Yin, D., Chen, Y., Ramchandran, K., & Bartlett, P. (2018). Byzantine-Robust Distributed Learning: Towards Optimal Statistical Rates. ArXiv, abs/1803.01498.

[74] Vepakomma, P., Swedish, T., Raskar, R., Gupta, O., & Dubey, A. (2018). No Peek: A Survey of Private Distributed Deep Learning. ArXiv, abs/1812.03288.

[75] Wei, K., Li, J., Ma, C., Ding, M., & Poor, H. (2021). Differentially Private Federated Learning: Algorithm, Analysis and Optimization. In: *Studies in Computational Intelligence*, Vol. 965, 51–78. Springer Science and Business Media Deutschland GmbH. https://doi.org/10.1007/978-3-030-70604-3_3

[76] Mohassel, P., & Rindal, P. (2018). ABY3: A Mixed Protocol Framework for Machine Learning. *Proceedings of the 2018 ACM SIGSAC Conference on Computer and Communications Security*, October 2018, 35–52. https://doi.org/10.1145/3243734.3243760

[77] Reina, G.A., Gruzdev, A., Foley, P., Perepelkina, O., Sharma, M., Davidyuk, I., Trushkin, I., Radionov, M., Mokrov, A., Agapov, D., Martin, J., Edwards, B., Sheller, MJ, Pati, S., Moorthy, P.N., Wang, S., Shah, P., & Bakas, S. (2021). OpenFL: An Open-Source Framework for Federated Learning. ArXiv, abs/2105.06413.

[78] Lai, F., Dai, Y., Zhu, X., & Chowdhury, M. (2021). FedScale: Benchmarking Model and System Performance of Federated Learning. ArXiv, abs/2105.11367.

[79] Yang, Q., Liu, Y., Cheng, Y., Kang, Y., Chen, T., & Yu, H. (2019). Federated Learning. Synthesis Lectures on Artificial Intelligence and Machine Learning. In: *Federated Learning. Synthesis Lectures on Artificial Intelligence and Machine Learning*. Springer. https://doi.org/10.1007/978-3-031-01585-4_6

[80] Lyu, L., Yu, J., Nandakumar, K., Li, Y., Ma, X., Jin, J., Yu, H., & Ng, K.S. (2020). Towards Fair and Privacy-Preserving Federated Deep Models. *IEEE Transactions on Parallel and Distributed Systems, 31*, 2524–2541.

[81] Lyu, L., Xu, X., & Wang, Q. (2020). Collaborative Fairness in Federated Learning. ArXiv, abs/2008.12161.

Chapter 2

Utility-based recommendation system for large datasets using EAHUIM

Vandna Dahiya
Maharshi Dayanand University

2.1 INTRODUCTION

To boost user engagement and increase purchasing potential, a growing number of online businesses are resorting to recommendation systems (RSs), and this trend is set to continue [1,2]. RSs [3] (also known as "recommendation engines") can change the way websites communicate with their visitors and benefit businesses by maximizing the return on investment by accumulating data on each customer's preferences and purchases. Companies like Amazon, Myntra, and Flipkart deploy RS as an important tool to create a personalized user experience. Customers' demographic data is collected, analyzed, and linked to their shopping experience, such as purchases, ratings, and so on. The inferences are then used to forecast the likelihood of similar products in the future. According to McKinsey & Company, Amazon's RS accounts for 35% of the company's profitability. For a brand with tens of thousands of products, hard-coding product suggestions for all of them would be challenging, and such static suggestions would quickly become out-of-date or obsolete for many customers. E-Commerce powerhouses can employ a variety of "filtering" approaches to discover the optimal times to suggest new things we are willing to buy (via their website, by email, or in any other way) [4]. Recommendation engines not only help customers find more of what they need, but they also enhance cart value. It is also worth mentioning that RS is more likely to be a suitable fit for businesses with many data and AI technologies in-house.

There are very few approaches in the literature that address the issue of revenue optimization through personalized product recommendations [5,6]. Whereas [5] concentrates on static circumstances that are ineffective in real-world big data because of its fast-paced nature, [6] typically requires a great deal of prior knowledge, such as consumer adoption levels over time for each product. Furthermore, only a few existing approaches can process streaming data to make decisions for the RS.

To overcome these issues, we concentrate on the real-time utility system for revenue management that can portray the information by optimizing the user's preferences. This chapter presents a real-time personalized

DOI: 10.1201/9781003244332-2

recommendation approach for e-Commerce using the High-Utility Itemset Mining (HUIM) technique. This method makes recommendations based on the users' preferences. To mine things with higher relevance to the user, researchers use "Enhanced Absolute High-Utility Itemset Mining" (EAHUIM) [7], a very efficient HUIM algorithm that has been proposed recently.

The significant difference between the work proposed in this chapter and the earlier studies is the integration of revenue management with user preferences on static and dynamic data.

The work is structured as follows: Section 2.2 emphasizes related research work. The problem statement is also defined here. The proposed approach is discussed in Section 2.3. Section 2.4 evaluates the model following the conclusion.

2.2 RELATED WORK

2.2.1 Recommendation system

RSs are information filtering systems capable of analyzing previous behavior and suggesting relevant items to the user. In other words, RS aims to guess the user's thoughts based on his previous behavior or that of other similar users in order to find the best scenario that matches the user's preferences. By delivering a personalized purchasing experience for each customer, RSs are a practical kind of targeted marketing [8]. There are different types of RS, including content-based RS, knowledge-based RS, and collaborative filtering-based RS. Based on the need, a suitable RS can be deployed. A good recommendation algorithm for large retailers should be scalable over large customer bases and product catalogs and require little processing time to generate online recommendations. It should also be able to respond swiftly to any change in users' data and offer appealing recommendations, regardless of how many purchases and ratings they have. Unlike other strategies, item-to-item collaborative filtering-based RS can mitigate these issues. To make a recommendation, the collaborative filtering method analyzes data from users. As a result, collaborative RS gather data based on users' preferences. Collaboration filtering based on association rules identifies relationships between individual items in a database, whereby the existence of one object in a transaction implies the presence of another object in the very same transaction.

RS applications exist in diverse areas such as medicine, agriculture, transportation, e-commerce, etc. [8]. They can make recommendations in various categories, including location-based information, movies, music, photos, books, etc. Users can save time and money using location-based data to locate themselves or predict their next position [9]. Furthermore, numerous stakeholders, such as farmers [10–12], healthcare personnel

[13–15], and tourists [16–18], might benefit from help in their decision-making strategy. The research used health RSs (HRSs) for dietary, exercise aid, and educational objectives. In this research, RS in e-commerce is considered to portray the model.

2.2.2 Recommendation systems in E-commerce

RSs are used to provide personalized product recommendations to website visitors. They learn from customers and recommend products that are relevant to them. Giant retailers such as Amazon and Flipkart rely on recommendation algorithms as vital and effective tools.

Figure 2.1 demonstrates various recommendations grounded in the contents of the shopping cart. Figure 2.2 demonstrates various recommendations made by an online medical application to the author based on his past shopping history.

E-commerce companies can benefit from RSs in three ways:

- Converting viewers to buyers: Most customers in an e-commerce store browse the items without purchasing, but if the site provides appropriate recommendations, they are more inclined to do so.
- Cross-sell: In addition to the items users are currently purchasing, RS suggests other goods to them. As a result, the average order size is expected to increase over time.
- Engagement: Loyalty is crucial in an age when an opponent's website may be accessed with only a few clicks. The user-site relationship is strengthened by RSs' personalizing the site for each user.

Customers who bought items in your cart also bought

Johnson's Baby Powder 600gm
★★★★☆ 210
₹247⁰⁰ ₹330.00
See buying options

Johnson's Baby Oil with Vitamin E (500ml)
★★★★☆ 574
₹352⁰⁰ ₹426.00
Add to Cart

Pampers Active Baby Taped Diapers, Extra Large size diapers, (XL) 56 c...
★★★★☆ 44,150
₹1,300⁰⁰ ₹1,649.00
Add to Cart

Dabur Baby Oil: Non - Sticky Baby Massage Oil with No Harmful Che...
★★★★☆ 231
₹227⁰⁰ ₹475.00
See buying options

Figure 2.1 Recommendations based on the contents of the shopping cart.

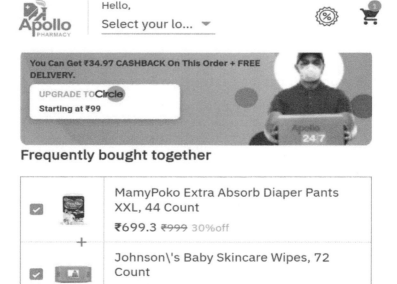

Figure 2.2 Example of recommendations suggested to the user by an e-medical application.

2.2.3 Recommendation system with utility itemset mining

An online e-commerce platform would like to show recommendations to a current customer to maximize the likelihood of the customer making an additional purchase during the current transaction. This necessitates communication between the active customer, the e-commerce website, and all previous e-commerce website transactions. Typically, an e-commerce store's transaction dataset is enormous. As a result, the RS should promptly provide appropriate recommendations to the active client while also being scalable. In online shopping, a product's reputation is decided by the feedback supplied by customers who have already purchased and used it. This information is typically collected in the form of user ratings, in which customers are given the option of assigning points to a product they have previously

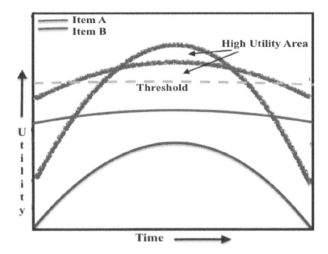

Figure 2.3 Utility distributions of two items over time.

purchased based on its usefulness. Other users look at the average ratings of similar products and try to make judgments based on them.

RSs also use the ratings offered by different users to anticipate which product(s) they might be interested in. Also, the majority of RS uses static datasets. A decent RS should be able to work with both static and dynamic data. The dynamic dataset is made up of variable transactions with varying traffic flows depending on the time of day and year. Seasonal peaks and trenches are possible. Figure 2.3 shows how the utility of the two items has changed throughout time. The demand for a product, and hence its utility, might fluctuate. It can also be demonstrated from the figure that even though a product's subjective utility value (item A) is lower than that of another product (item B), the overall utility can be higher over time.

2.2.4 Prerequisites of utility itemset mining

Some formal descriptions related to utility mining are quantified in this section, as they are the foundation of the proposed model.

I is the set of discrete items in the database.

$$I = \{I_1, I_2, I_3 \dots I_n\} \tag{2.1}$$

D is the dataset composed of transactions T, where T is a subset of I.

$$D = \{T_1, T_2, T_3 \dots T_m\} \tag{2.2}$$

F is the frequency or internal utility of the item in transaction T_t and is denoted as:

$$F(I_i, T_t)$$ (2.3)

S is a subjective value which denotes the significance of an item that is assigned by the user to express fondness and is denoted as:

$$S(I_i)$$ (2.4)

U is the utility function expressed as the product of internal utility and subjective value.

$$U(I_i, T_t) = F(I_i, T_t) * S(I_i)$$ (2.5)

X is an itemset of length k, and $1 \leq k \leq n$ represents the utility of itemset X in a transaction T_t.

$$U(X, T_t) = \sum U(I_i, T_t)$$ (2.6)

for $I_i \in X$ and $X = \{I_1, I_2...I_k\}$.
    U(X) is the utility of an itemset X in the database D.

$$U(X) = \sum U(X, T_t)$$ (2.7)

for $T_t \in D$ and $X \in T_r$.
    High-Utility Itemset (HUI)

$$HUI = \left\{ X \frac{1}{2} U(X)^3 \omega \right\}$$ (2.8)

where ω is the minimum utility threshold specified by the user.
    A dataset composed of various users' transactions can be seen in Table 2.1. Here, the user's shopping history US={US_1, US_2, US_3, ... US_U} is taken as the transactions, and items A, B, C, D, and E as the items of the dataset. Table 2.2 displays the subjective utility of the items.

Table 2.1 Transactional dataset for user's shopping cart

User	Item A	Item B	Item C	Item D	Item E
US_1	1	0	2	0	1
US_2	0	0	3	5	1
US_3	3	0	1	2	1
US_4	1	2	0	1	0
US_5	2	1	1	1	0
US_6	2	1	3	0	1

Table 2.2 Subjective value of items

Item	Subjective value
"A"	2
"B"	3
"C"	5
"D"	I
"E"	2

Now, an itemset is termed "High-Utility Itemset," if the utility is equal to or higher than the predefined threshold (assume 45).

Here, $U(A, US_1) = 1*2 = 2$

$U(B, US_1) = 0*2 =$

$U(C, US_1) = 2*5 = 10$ and so on.

Utility for an itemset is the sum of the utility of items. So, $U(\{A, C, E\}, US_1) = 2 + 10 + 2 = 14$.

Utility of itemset $U(\{A, C, E\})$ in the complete dataset is the sum of utility from all the transactions having this item set.

$$U(\{A, C, E\}) = U(\{A, C, E\}, US_1) + U(\{A, C, E\}, US_3)$$

$$+ U(\{A, C, E\}, US_6)$$

$= 14 + 13 + 21 = 48$, which can be termed as a HUI.

HUIM aims to discover all sets of items or itemsets with a utility value equal to or greater than the predefined threshold.

2.2.5 Problem definition

The goal is to create a scalable product RS in which users add things to their basket and then receive recommendations based on the products in the cart. The RS should provide the best possible recommendations from the database (static and dynamic) that will increase the likelihood of the user purchasing an additional product, and the recommendations should be displayed with the shortest feasible delay.

2.3 PROPOSED MODEL

To address the challenge of real-time suggestions using association rules, a new framework, "Utility-Based Recommendation System" is proposed, which is a personalized RS that utilizes the EAHUIM [7,19,20] technique. The EAHUIM is a technique for mining itemsets with high utility. EAHUIM is a scalable algorithm with a distributed structure to mine HUIs from large datasets. The algorithm scales effortlessly with an increase in data size. The model works in two phases. The first phase consists of extracting high-utility itemsets using the EAHUIM algorithm. The second phase of the model consists of recommendations based on the antecedents as provided by the user. An adaptive approach is also framed for load management if the data is not static. The design of the model is explained in detail in this section.

2.3.1 Design of the model – recommendation system with EAHUIM

As mentioned above, the RS with EAHUIM or RS-EAHUIM works in two phases. These phases are described in the following sections.

2.3.1.1 Phase I – Mining HUIs using EAHUIM

The first phase of the model consists of framing association or correlation rules for various variables/items in the data. The data for recommendations is selected, preprocessed, and then imported into an HDFS file in the case of static data. The algorithm EAHUIM is used to mine the patterns/itemsets with high utility when they occur together. EAHUIM is distributed in nature and scales well with increased data size. A utility threshold value is specified externally and used as the minimum criteria for mining HUIs. Data is divided into blocks of equal size, and "transaction weighted utilities" (TWU) for the items are computed using the map-reduce framework. Items with TWU values greater than the threshold are only carried further in the modified database. This modified dataset is then partitioned among various workers/nodes using a neutral load balancing technique so that a node only stores the data it requires for further computations. This way, the storage complexity is reduced to a great extent. Various pruning methods are employed to avoid unnecessary computations. Every node is responsible for mining itemsets from the data given to it. Complete sets of HUIs are collected at the end of the algorithm, which serves as the associations' rules for the given data. The process for phase I is depicted in Figure 2.4.

A dynamic strategy to estimate utility values in streaming data is being proposed. A load factor Θ (ranging from 0 to 1) is used to measure the flow of transactions, distinguishing between light and heavy flows, and the model is adapted accordingly. The fluctuating flow of transactions is shown in Figure 2.5. The model is switched between a single system approach and a distributed approach based on the load factor, as shown in equation 2.9.

$$Q = 1/1 + e^{-\alpha(N-\alpha)} \tag{2.9}$$

where N is the number of transactions during a given time period, and α is the control parameter.

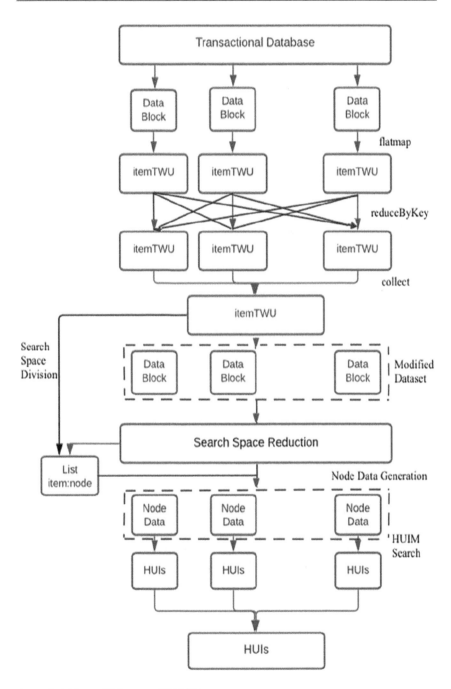

Figure 2.4 Mining HUIs using EAHUIM.

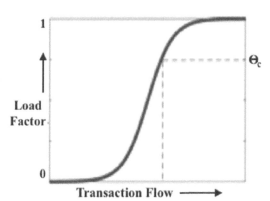

Figure 2.5 Transaction flow curve.

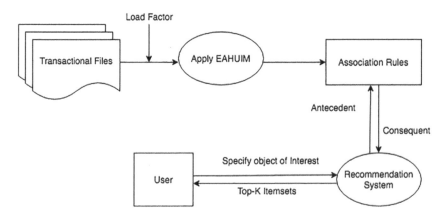

Figure 2.6 Design of the proposed model.

2.3.1.2 Phase II – Recommendation to the user

The EAHUIM algorithm's output is saved in the form of association rules. When a user interacts with the system and searches for an item, the antecedents work as input items, and the top-most consequent items from the associations are suggested to the user. The design of the suggested model is depicted in Figure 2.6. An example of the antecedent and consequent of the association rule is shown below:

$Item_1 \rightarrow item_3$
$Item_2, item_4 \rightarrow item_6$
$Item_2 \rightarrow item_3$

So, whenever an item is searched for or entered in the cart/basket, the model starts finding the corresponding high-utility items recommended to the user, as shown in Figures 2.1 and 2.2.

Two variants are being proposed for RS using EAHUIM (RS-EAHUIM) to maximize the user's personalized experience. RS-EAHUIM-F is an item-frequency-based approach that endorses the item based on its occurrence frequency. Here, the subjective utility is taken as the same (let 1) for all the items. So "utility" is defined as a frequency or internal utility only. RS-EAHUIM-R is an item-ratings-based approach. The users provide ratings based on their shopping experiences, which are generally in numerical form. Although high ratings normally imply a high degree of customer adoption and liking for specific things, a low rating does not imply that a product will not be purchased. Users may purchase an item and then give it a poor rating because they dislike it. An item's average rating, or the interestingness factor, is taken as subjective utility value. Hence, the utility function of an item can be computed as a product of *internal utility* and *average_rating*.

Taking the example of the transactional dataset from Tables 2.1 and 2.2, which represent the shopping cart from an e-commerce store and the subjective values of the items, the utility of the itemsets can be computed as follows:

$$U(\{A\}, US_1) = 2$$

$$U(\{C\}, US_1) = 10$$

$$U(\{A, C\} = U(\{A, C\}, US_1) + U(\{A, C\}, US_2)$$
$$+ U(\{A, C\}, US_5) + U(\{A, C\}, US_6)" = 51$$

If the utility threshold is set as 45, itemset {A, C} is a HUI and can be used in recommendations. However, if the frequency threshold is set as 6, {A, C} cannot be considered an interesting itemset. For the thresholds mentioned above, {E} is a frequently-cited item but not a HUI as U{E}=9, which does not satisfy the user's favorites.

The performance of these two variants is evaluated in the following section.

2.4 RESULTS AND DISCUSSION

Extensive experiments are being performed to evaluate the proposed approach using a real-world dataset from an e-commerce website [21]. The users' ratings of products are considered subjective utility values for the items.

2.4.1 Setup

The model "Utility-Based Recommendation System" has been set up with a Spark cluster having five nodes, where one node is a master node and the other four nodes act as slave nodes. The master node features 8 GB of RAM and a 2.80 GHz Intel® CoreTM i7 processor. Each worker node has 4 GB of RAM. The following are the software settings on each node: Windows 10, Spyder4, and Spark version 3.0. Python is used for programming. To assess the competence of the proposed system, three different variants are used, as shown in Figure 2.7.

2.4.1.1 RS-EAHUIM-F

As stated earlier, RS-EAHUIM-F is an item-frequency-based RS employing the algorithm EAHUIM, where subjective utility is the same for all the items and only the occurrence frequency of items is considered in the utility function. Here, subjective utility is considered "1" for the experimental purpose.

2.4.1.2 RS-EAHUIM-R

RS-EAHUIM-R is an EAHUIM-based RS where occurrence frequency and subjective utility are considered in the utility function. Thus, the utility function is composed of the product of internal utility (frequency) and subjective utility (utility factor). Here, the product's average rating is taken as subjective utility.

2.4.1.3 RS-Two-Phase-R

A third variant of the RS is considered that employs the Two-Phase standard algorithm for mining itemsets with a utility function. Here, the utility function is also composed of the product of internal utility (frequency) and subjective utility (utility factor). The average rating of the product is taken as subjective utility. This model is taken as the ground truth for the system.

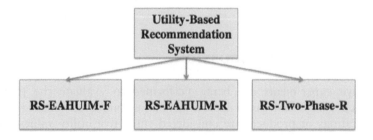

Figure 2.7 Different variants of recommendation systems.

2.4.2 Data collection and preprocessing

The Amazon [21] dataset for electronic items has been used to evaluate the model. The dataset has been reviewed in text and ratings and is composed of item metadata that contains descriptions, cost, make, and other information for an item. The dataset is composed of 20,994,353 ratings by 4,201,696 unique users. Table 2.3 provides an instance of the dataset.

For the performance evaluation of this model, only three relevant columns have been chosen – UserId, ProductId, and Ratings. Every person and product has a unique identification, which is presented as UserId and ProductId, and the ratings show the numerical value of the product's likelihood. The extracted dataset is composed of 1,048,576 ratings, 786,330 unique userIds, and 61,894 productIds, where every user and item has at least 5 ratings. These statistics have been taken to reduce the sparsity of the data. The average rating of a product is computed using the *group by* function. The dataset is converted into a transactional form suitable for the proposed model, where each user is considered a transaction and the products are considered items in the database. Table 2.4 reveals the data types for the attributes of the datasets. As the data is static and large, the load factor is considered heavy, and the evaluation is done within the distributed framework of the algorithm.

2.4.3 Performance evaluation

To assess the model's performance, item B0057HG70H is selected randomly as the antecedent and fed to the model. The top ten outcomes of the three variants are displayed in Table 2.5. The following parameters are used for assessment.

Table 2.3 Amazon-electronics dataset snippet (only three columns)

UserId	ProductId	Ratings
A1QGNMC6O1VW39	0594017580	2.0
A28B1G1MSJ6OO1	0511189877	3.0
AMO214LNFCEI4	0594012015	5.0
A1H8PY3QHMQQA0	1397433600	1.0
A3J3BRHTDRFJ2G	0528881469	2.0
A265MKAR2WEH3Y	0528881469	4.0

Table 2.4 Data types of attributes

S/N	Attribute	Data type
1	UserId	Object
2	ProductId	Object
3	Rating	float64

Table 2.5 Top ten items obtained from three experiments

Rank	RS-EAHUIM-F	RS-EAHUIM-R	RS-TwoPhase-R
I	1400501776	1400501776	1400501776
2	B0002L5R78	B0002L5R78	B0002L5R78
3	B000JYWQ68	9984984354	9984984354
4	B0002L5R78	B0002L5R78	B0002L5R78
5	1400532620	9984984354	9984984354
6	G977RDH70I	07GVI20GYI	07GVI20GYI
7	086XBM36D8	0KTDBJI580	0KTDBJI580
8	0I9HY75DH6	AJUIBH8I20	AJUIBH8I20
9	07GVI20GYI	A864V02G76	A864V02G76
I0	0KTDBJI580	03VJ78990V	03VJ78990V

2.4.3.1 *Accuracy of the model*

The variants display different results for the consequents. The top ten recommended items are displayed in Table 2.5. The top ten items suggested by RS-EAHUIM-F are different from the recommendations of RS-EAHUIM-R and RS-Two-Phase-R. The variations are shown in bold letters. For example, item {B000JYWQ68} is recommended by item-frequency-based RS but not by rating-based RS. The utility of this item, along with the user-specified item, is counted manually using the *countby* function, which was found to be lower than the itemset placed in third position in ratings-based RS. Ratings based on RS make room for items with more ratings or preferences.

The top ten recommendations from ratings-based experiments are the same, which illustrates the accuracy of the proposed model RS-EAHUIM-R concerning the Two-Phase algorithm. Although the general accuracy of EAHUIM is 94.6% compared to the Two-Phase algorithm, the former faces a trade-off using various pruning strategies to speed up the algorithm [22]. But, here, only the top ten recommendations are considered, which are the same for both approaches.

2.4.3.2 *Quality of recommendations*

To see the quality of recommendations, the occurrence frequency of the top ten itemsets is being compared, as shown in Table 2.6. The average occurrence frequency for the top ten items for RS-EAHUIM-F is 282 (rounded off), whereas for RS-EAHUIM-R and RS-Two-Phase, it is 251 (rounded off). That shows a significant improvement in the recommendations. The above investigations have shown that a utility-based recommendation method is effective at meeting a user's personal preferences. Figure 2.8 shows the occurrence frequencies of the top ten items suggested by the ratings-based

Table 2.6 Occurrence frequency of top ten itemsets

Rank	1	2	3	4	5	6	7	8	9	10
Frequency-based RS	382	341	319	298	274	265	249	243	238	212
Ratings-based RS	382	341	269	298	243	238	212	160	178	189

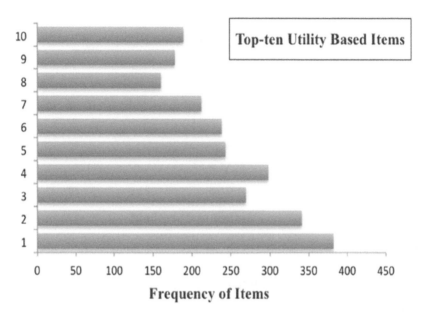

Figure 2.8 Occurrence frequencies of top ten itemsets.

RS. There is only a small but significant deviation from the frequency-based RS, demonstrating the model's usefulness. The results validate that the items are suggested based on the earlier preferences of the user and provide a more personalized approach than item-frequency-based systems.

2.4.3.3 Execution time

The time complexity of variants RS-EAHUIM-F and RS-EAHUIM-R is the same as they employ the same EAHUIM algorithm. However, they are much faster than RS-Two-Phase-R, which employs the non-distributed Two-Phase algorithm. Table 2.7 shows the results for the execution time of the EAHUIM-based RS and the Two-Phase based RS. Figure 2.9 shows that EAHUIM-based approaches are much faster and provide near-real-time recommendations for all the values of rating thresholds.

Table 2.7 Scalability of the model RS-EAHUIM

Number of working nodes	I	2	3	4	5
Execution time (seconds)	795	618	423	326	224

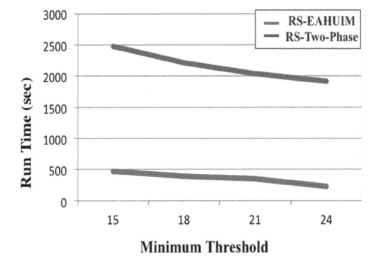

Figure 2.9 Run time performances of RS with EAHUIM and Two-Phase.

2.4.3.4 Scalability of the model

The pool of working nodes is changed to examine the adaptability of the RS-EAHUIM to the rating model. The execution time with different nodes can be seen in Table 2.7. The speedup has been proven to be almost linear, meaning that the model with the algorithm EAHUIM would scale adequately as the count of nodes increases. Nonetheless, at a definite number of processors, the costs of synchronization and communication tend to outweigh the costs of processing overhead, and introducing additional processors has little influence on computation time. Obtaining a linear speedup is unrealistic, considering communication costs will dominate overall operating time. Scalability is depicted in Figure 2.10 by adjusting the available nodes from 1 to 5 while running at a 24% threshold.

2.4.3.5 Stability of the model

The proposed model, RS-EAHUIM, remains stable for different thresholds and does not go out of bounds. This validates the stability of the model.

The models are not comparable in terms of memory requirements as they have different structures for the algorithms. An EAHUIM-based RS utilizes the memory from all the available working nodes, whereas Two-Phase-based

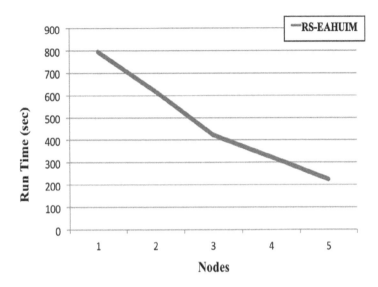

Figure 2.10 Execution time of the model RS-EAHUIM with varying nodes.

RSs can only use the memory available for the stand-alone system, being Two-Phase, a centralized algorithm.

2.4.4 Limitations of the model

The collaborative filtering-based RS suffers from two significant problems – cold start and data sparsity [23]. The cold-start issue happens when an item or a user is introduced to the RS for the first time. Data sparsity arises when the RS emphasizes personalization, and there are few or no ratings/subjective utility values available for most items. Various similarity-based methods have been proposed and applied to find similarities between users and predict the ratings for unrated items to tackle the abovementioned problems. However, they are outside the scope of this study and can be included in future work.

2.4.5 Discussions

Several organizations are exploring the potential of big data analytics [24–32]. The objective is to extract meaningful insights from massive datasets, allowing stakeholders to draw inferences and work accordingly. It is also imperative to ensure such outcomes are implemented in real time. The model suggested in this study utilizes the distributed HUI mining method EAHUIM, which maintains a balance between the frequency and utility of database items and gives suggestions in near real-time. The work may be used to build several information management and data-driven decision-making systems.

2.5 CONCLUSION

By delivering a tailored experience, RSs present an effective form of targeted marketing. For large retailers, a good recommendation algorithm should be scalable and deliver recommendations in the shortest time possible. Even though RSs have long been used in industry and academia, there are still challenges in this domain. A load-adaptive utility-based recommendation model has been proposed in this chapter using the EAHUIM algorithm, which provides a balance between frequency and interestingness factors. Experiments performed on e-commerce data demonstrate that the utility approach used in this chapter provides near-real-time and personalized recommendations, which proves the model's efficiency.

REFERENCES

[1] Fayyaz, Z., Ebrahimian, M., Nawara, D., Ibrahim, A., & Kashef, R. "Recommendation systems: algorithms, challenges, metrics, and business opportunities", *Applied Sciences*, vol. 10, Art. no. 2, 7748, pp. 1–20, 2020, https://doi.org/10.3390/app10217748

[2] Aggarwal, C.C. "An introduction to recommender systems", in *Recommender Systems*. Springer: Cham, Switzerland, 2016, pp. 1–28, https://doi.org/10.1007/978-3-319-29659-3_1

[3] Ricci, F., Rokach, L., & Shapira, B. "Recommender systems: introduction and challenges", in *Recommender Systems Handbook*. Springer: Boston, MA, USA, 2015, pp. 1–34, https://doi.org/10.1007/978-1-4899-7637-1_1

[4] Linden, G., Smith, B., & York, J. "Amazon.com recommendations: item-to-item collaborative filtering", *IEEE Internet Computing*, vol. 7, pp. 76–80, 2003, https://doi.org/10.1109/MIC.2003.1167344

[5] Belkin, N.J. & Croft, W.B. "Information filtering and information retrieval: two sides of the same coin", *Communications of the ACM*, vol. 35, (12), pp. 29–38, 1992, https://doi.org/10.1145/138859.138861

[6] Joaquin, D. & Naohiro, I. "Memory-based weighted-majority prediction", *Annual International ACM SIGIR Workshop on Recommender Systems: Algorithms and Evaluation*, 1999.

[7] Dahiya, V., & Dalal, S. "Enhanced absolute high utility itemset miner for big data", *Journal of Information Management Data Insights*, vol. 1, (2), Article No. 100055, 2021.

[8] Srifi, M., Oussous, A., Ait Lahcen, A., & Mouline, S. "Recommender systems based on collaborative filtering using review texts: a survey", *Information*, vol. 11, (6), pp. 317, 2020, https://doi.org/10.3390/info11060317

[9] Rodriguez-Carrion, A., Garcia-Rubio, C., Campo, C., Cortés-Martín, A., Garcia-Lozano, E., & Noriega-Vivas, P. "Study of LZ-based location prediction and its application to transportation recommender systems", *Sensors*, vol. 12, (6), pp. 7496–7517, https://doi.org/10.3390/s120607496

[10] Madhusree, K., Rath, B.K., & Mohanty, S.N. "Crop recommender system for the farmers using Mamdani fuzzy inference model", *International Journal of Engineering and Technology*, vol. 7, (4), pp. 277–280, 2018, https://doi.org/10.14419/ijet.v7i4.15.23006

[11] Jaiswal, S., Kharade, T., Kotambe, N., & Shinde, S. "Collaborative recommendation system for agriculture sector", *ITM Web of Conferences*, vol. 32, pp. 03034, 2020, https://doi.org/10.1051/itmconf/20203203034

[12] Bangaru Kamatchi, R.S. "Improvement of crop production using recommender system by weather forecasts", *Procedia Computer Science*, vol. 165, pp. 724–732, 2019, https://doi.org/10.1016/j.procs.2020.01.023

[13] Sezgin, E., & Sevgi, Ö. "A systematic literature review on health recommender systems", in *Proceedings of the 2013 E-Health and Bioengineering Conference (EHB)*, Iasi, Romania, IEEE: Piscataway, NJ, USA, 2013, pp. 1–4, https://doi.org/10.1109/EHB.2013.6707249

[14] Sung, Y.S., Dravenstott, R.W., Darer, J.D., Devapriya, P.D., & Kumara, S. "SuperOrder: provider order recommendation system for outpatient clinics", *Health Informatics Journal*, vol. 26, pp. 999–1016, 2019, https://doi.org/10.1177/1460458219857383

[15] Sahoo, A.K., Pradhan, C., Barik, R.K., & Dubey, H. "DeepReco: deep learning based health recommender system using collaborative filtering", *Computation*, vol. 7, (2), pp. 25, 2019, https://doi.org/10.3390/computation7020025

[16] Sperlì, G., Amato, F., Mercorio, F., & Mezzanzanica, M. "A social media recommender system", *International Journal of Multimedia Data Engineering & Management*, 2018, vol. 9, (1), pp. 36–50, RePEc:igg:jmdem0:v:9:y:2018:i:1:p:36-50

[17] Mwinyi, I.H., Narman, H.S., Fang, K., & Yoo, W. "Predictive self-learning content recommendation system for multimedia contents", in *Proceedings of the 2018 Wireless Telecommunications Symposium (WTS)*, Phoenix, AZ, USA, IEEE: Piscataway, NJ, USA, 2018, pp. 1–6, https://doi.org/10.1109/WTS.2018.8363949

[18] Zhang, M., Yi, C., Xiaohong, Z., & Junyong, L. "Study on the recommendation technology for tourism information service", in *Proceedings of the Second International Symposium on Computational Intelligence and Design, 2009*; IEEE: Piscataway, NJ, USA, 2009, vol. 1, pp. 410–415.

[19] Dalal, S., & Dahiya, V. "Performance comparison of absolute high utility itemset mining (AHUIM) algorithm for big data", *International Journal of Engineering Trends and Technology*, vol. 69, (1), pp. 17–23, 2021, https://doi.org/10.14445/22315381/IJETT-V69I1P203

[20] Dalal, S., & Dahiya, V. "A novel technique - absolute high utility itemset mining (ahuim) algorithm for big data", *International Journal of Advanced Trends in Computer Science and Engineering*, vol. 9, (5), pp. 7451–7460, 2020, https://doi.org/10.30534/ijatcse/2020/78952020

[21] Ni, J., Li, J., & McAuley, J. "Justifying recommendations using distantly-labeled reviews and fine-grained aspects", *Empirical Methods in Natural Language Processing, China*, pp. 188–197, 2019, https://doi.org/10.18653/v1/D19-1018; https://nijianmo.github.io/amazon/index.html

[22] Dahiya, V., & Dalal, S. "A scalable approach for data mining - AHUIM", *Webology*, vol. 18, (1), pp. 92–103, 2021, https://doi.org/10.14704/WEB/V18I1/WEB18029

[23] Al-Bakri, N.F., & Hashim, S.H. "Reducing data sparsity in recommender systems", *Journal of Al Nahrain University*. 2018, vol. 21, (2), pp. 138–147, https://doi.org/10.22401/JNUS.21.2.20

[24] Yang, R., Xu, M., & Nagiza, S. "Real-time utility-based recommendation for revenue optimization via an adaptive online top-k utility itemsets mining model", *13th International Conference on Natural Computation, Fuzzy Systems and Knowledge Discovery (ICNC-FSKD)*, 2017, https://doi.org/10.1109/FSKD.2017.8393050

[25] Bennett, J., & Lanning, S. "The Netflix prize." *Proceedings of KDD Cup and Workshop*, 2007.

[26] Errico, J.H., et al. "Collaborative recommendation system." US Patent No. 8,949,899, 3 February 2015.

[27] Liang, S., Liy, Y., Jian, L., & Gao, Y. "A utility-based recommendation approach for academic literatures", *ACM International Conferences on Web Intelligence and Intelligent Agent Technology*, vol. 3, pp. 229–232, 2021, https://doi.org/10.1109/WI-IAT.2011.110

[28] Dhanda, M., & Verma, V. "Personalized recommendation approach for academic literature using itemset mining technique", Springer Nature, *Progress in Intelligent Computing Techniques*, pp. 247–254, 2018, https://doi.org/10.1007/978-981-10-3376-6_27

[29] Sandeep, D., & Vandna, D. "Review of high utility itemset mining algorithms for big data", *Journal of Advanced Research in Dynamical and Control Systems*, vol. 10, (4), pp. 274–283, 2018.

[30] Reddy, S., Ranjith, & Divya, K.V. "Association rule-based recommendation system using big data", *International Journal of Engineering and Technology*, vol. 4, (6), pp. 652–657, 2017.

[31] Beel, J., Gipp, B., Langer, S., & Breitinger, C. "Research-paper recommendation systems: a literature survey", *International Journal on Digital Libraries*, vol. 17, (4), pp. 305–338, 2016, https://doi.org/10.1007/s00799-015-0156-0

[32] Moscato, V., Picariello, A., & Sperli, G. "An emotional recommender system for music", *IEEE Intelligent Systems*, vol. 36, pp. 57–68, 2021.

Chapter 3

Anaphora resolution
A complete view with case study

Kalpana B. Khandale and C. Namrata Mahender
Dr. Babasaheb Ambedkar Marathwada University

3.1 INTRODUCTION

In an era when natural language processing (NLP) is gaining popularity, the problem of anaphora resolution is becoming increasingly important for various applications, including machine translation, information extraction, and question-answering. Due to the ambiguous character of natural languages, it is difficult for a machine to determine the correct antecedent to the anaphora. Anaphora is a primary discursive strategy used to reduce repetition and promote text cohesion by requiring the interpretation of subsequent sentences to be dependent on the interpretation of preceding sentences.

"Mahendra Singh Dhoni is a cricket player. In 2011, he was distinguished as the world's best player."

A human reader will immediately recognize that Mahendra Singh Dhoni was named the most outstanding player in 2011 in the second statement. This deduction, however, necessitates establishing a relationship between Mahendra Singh Dhoni in the first sentence and he in the second. Only then can Mahendra Singh Dhoni be credited with the award specified in the second sentence. As a result, the interpretation of the second sentence is dependent on the first, ensuring that the discourse's two sentences are coherent. The two expressions are co-referential since they refer to the same person in the real world, Mahendra Singh Dhoni, contributing to the discourse's cohesiveness.

3.1.1 Issues and challenges of an Anaphora

The computational complexity of a fundamental computational problem in human language, detecting references related to items in linguistic representation, is investigated in this chapter. Since antiquity, the question of how linguistic pieces find their reference has been a key theme in studying language and thought. However, this computer challenge has never been clearly characterized or its computational complexity examined.

DOI: 10.1201/9781003244332-3

Because other language elements do not mediate their relationship, some linguistic elements have what is known as *a direct reference*. Referring expressions are the name for these elements. Proper nouns (such as Swara and Viraj) and definite noun phrases are examples (for example, my friend, the woman dressed in pink). These elements pose a computational issue in that they must find a unique cognitive referent for each of them in a given linguistic representation. For example, to convey the phrase Salim and Anarkali, the language might associate the proper nouns Salim and Anarkali with his mental images of Salim and Anarkali.

Another linguistic element known as the anaphoric element is an *indirect reference*, which means their reference depends on other linguistic elements' references. Pronouns (he, her, she, him, it, they), reciprocals (each other), and reflexives (himself, themselves) are examples of anaphoric elements. Antecedents are required for anaphoric constituents. These elements are computationally difficult to locate a unique antecedent for each anaphoric element in a given language representation. This is one of the anaphora issues. To depict the utterance, *Neelam adores herself*; the language used must first decide that the proper noun Neelam is the intended antecedent of the reflexive herself and that hence herself refers to whomever Neelam is referring to.

The problem of an anaphora in computation is entirely based on language and produces an antecedence relation on linguistic elements independent of other cognitive systems. On the other hand, the referential interpretation issue defines a computation whose output links the language system to other cognitive systems.

As a result, the language user can appropriately assign an antecedent to any anaphoric element in a linguistic representation without determining the reference of any element in the representation. The following statement exemplifies this:

"She is very proud of her beauty, but she felt ashamed when she saw her image in the mirror."

The language user can quickly determine that the anaphoric elements (she) in this utterance all refer to the same person without knowing what that reference is. As a result, the issue of an anaphora can be solved even if the issue of referential interpretation remains unsolved. Furthermore, every anaphoric element in that representation cannot be assigned a reference. Therefore, an anaphora issue must be solved to solve an issue of referential interpretation.

NLP considers anaphora and co-reference resolution to be tough tasks. Anaphora resolution algorithms have not yet achieved accuracy high enough to be deployed in higher-level NLP applications, despite significant progress in anaphora resolution over the last few years [1]. This is due to the difficulties in resolving anaphora. As the introduction says, anaphora resolution is a discourse-level problem requiring various lower-level linguistic skills, including morph analysis, POS tagging, and syntactic and semantic structures. As a result, anaphora resolution algorithms must rely on NLP tools and analysis for preprocessing. As a result of the poor performance and mistakes in these preprocessing techniques, the anaphora resolution algorithms' accuracy suffers. While some of these preprocessing methods, such as POS tags, have been shown to be reasonably reliable, others, such as parsing, require significant refinement before being utilized for anaphora resolution.

Even if all the necessary linguistic knowledge is present in the corpus or through preprocessing techniques, statistical methodologies require an annotated corpus for training or learning. Building an annotated anaphora corpus has been the focus of many research endeavors. Much work has been done on identifying and utilizing numerous factors that serve as the foundation for anaphora resolution algorithms. However, defining the core set of characteristics employed for anaphora resolution is still a work to be explored.

In any NLP application, the treatment of anaphora is critical to comprehending and generating coherent discourse. The goal of the anaphora resolution issue is to discover the proper antecedent for a particular anaphora in discourse. Anaphora resolution and its integration in diverse NLP applications focus on the research discussed in this chapter. The task of resolving anaphora in English has been intensively researched, with various unique techniques and implementations. Other languages, such as Spanish, French, Italian, Japanese, Korean, and German, have some methods, but the NLP research done for these languages is not comparable to the study done for English.

The problem of an anaphora is a classification of computations carried out by a human language user. Each of these computations represents information about a language user's linguistic utterances. In the instance of anaphora, the language computation's output represents the language user's knowledge of the anaphoric elements in the utterance's antecedents. The computational theory of anaphora details how language users calculate their knowledge about anaphoric items. As a result, a complete statement of an issue of anaphora is a critical component of any empirically acceptable human language theory.

3.1.2 Need for anaphora in NLP applications

Anaphora resolution, in general, is inspired by its real-time applications in NLP. Some of the most important uses are discussed in the following sections.

3.1.2.1 Question-answering system

The task of question-answering entails finding the answer to a natural language inquiry from a given text. Most basic question-answering systems utilize simple string matching or a parse analysis of a sentence. An essential sub-task of the question-answering system is anaphora resolution.

"Mahatma Gandhi took the path of non-violence. He was a great person."

Given the above text and a query as "Who was great personality?", When we use string matching, the second sentence will respond "He," which is not a complete answer. As a result, to get the real answer, the system must know that the pronoun "He" in the prior sentence refers to a different entity, i.e., *Mahatma Gandhi.*

3.1.2.2 Text summarization

Automatic summarization is known as reducing a written document with computer software to generate a summary that preserves the most significant aspects of the original content. Though there are several ways and variations for text summarizing, the most frequent strategy is to choose only the most relevant information and discard the rest. Although summary is often employed for long texts, consider the below example to be summarized for simplicity. A possible summary of the text in the example may be generated by connecting the following sentences.

"Mahatma Gandhi took the path of non-violence, was a great person."

This, too, necessitates knowledge of the pronoun's referent. Hence, the anaphora resolution method is useful in summarization. Without resolving the anaphors, summarization might result in sentences containing anaphors but no obvious referent, rendering the summary unintelligible and nonsensical.

3.1.2.3 Machine translation

Machine translation translates text from one language to another using computers and software. Most machine translation research has concentrated on sentence-level translation to decrease the complexity of translation, i.e., translation of individual sentences while they are being considered separately. However, discourse-level information is required to get a meaningful translation of longer texts in real time. In certain situations, pronoun translation from one language to another necessitates knowledge of the pronouns' referents. This is because pronouns in certain languages take on multiple forms depending on their morphological characteristics, such as number, gender, and animacy of the word. As a result, knowing the exact translation of such a pronoun is required. The pronoun's referent, which comes via anaphora resolution, is necessary.

3.1.2.4 Sentiment analysis

This field helps find users' sentiments from a text like a review, comment, etc. This task describes the emotional connection behind a particularly positive, negative, or neutral text.

For a few applications discussed above, other than this, anaphora resolution is necessary for information extraction, dialogue systems, speech recognition, sentiment analysis, document retrieval, and many more.

3.1.3 Anaphora

Anaphora is the usage of an expression that relates to another term in its context in linguistics. The referring expression is anaphor, whereas the referred expression is known as the antecedent. The antecedent offers information for the anaphor's interpretation. Consider the following scenario.

"Ram is the bank manager, but he cannot break the law."

In the above example, "He" refers to a previously mentioned entity, "Ram." Determining the correct antecedent to the correct anaphora is called "anaphora resolution."

Anaphora can also be classified based on the placement of the antecedent: intrasentential if the antecedent is in the same sentence as the anaphora, or intersentential if the anaphoric relation is made across sentence boundaries. In addition to the extensive information required to resolve anaphora, the different shapes that anaphora can take make it a complex undertaking, especially when teaching computers how to solve anaphora.

1. Rahul is going to the market because he wants to purchase a new pair of shoes.
2. Rahul is going to the market. He is a good boy.

The examples above interpret the intrasentential (1) and intersentential (2) sentences.

3.1.3.1 Types of anaphora

There are various types of anaphora.

3.1.3.1.1 Pronominal anaphora

Pronominal anaphora occurs once at the level of

- **Personal** (he, him, she, her, it, they, them)
  For example, *Ravi needed to get ready for a meeting, so he shaved.*
- **Possessive** (his, her, hers, their, theirs)
  For example, *Ravi took out his old razor.*
- **Reflexive** (himself, herself, itself, themselves)
  For example, *However, he cut himself during shaving.*
- **Demonstrative** (this, that, these, those)
  For example, *This slowed him down even more. Ravi opted for the electric shaver. That did not work either.*

3.1.3.1.2 Lexical noun phrase anaphora

Syntactically, lexical noun phrase anaphora forms definite noun phrases, also known as definite descriptions and proper names.

For example, *Mahesh took a drive and became disoriented. That is Mahesh for you.*

3.1.3.1.3 Noun anaphora

The anaphoric link between a non-lexical preform and the head noun or nominal group of a noun phrase is not be confused with noun anaphora. A specific type of identity-of-sense anaphora is noun anaphora.

For example, *I will not have a sweet pretzel, just a plain one.*

3.1.3.1.4 Zero anaphora

The so-called zero anaphora or ellipsis is a type of anaphor based on the form of the anaphor. An anaphor is not visible because a word or phrase does not overtly convey it.

For example, *Seeta is exhausted. (She) had been working nonstop all day.*

3.1.3.1.5 One anaphora

It is a form of anaphora that refers to the preceding antecedent. Each has significance, but pronominal anaphora is the most common method for resolving anaphora.

For example, *I broke my phone and will have to get a new one.*

3.1.3.1.6 Nominal anaphora

When a referring expression (pronoun, definite noun phrase, or proper name) includes a non-pronominal noun phrase as its antecedent, nominal anaphora occurs.

3.1.3.1.7 Verb or adverb anaphora

The anaphora of a verb or adverb is determined by the anaphoric relationship between it and its antecedent in the preceding clause.

3.1.4 Discourse anaphora

The term "discourse" refers to the interconnection of sentences. Two or more sentences follow the initial initiative sentence. To correctly resolve the anaphora, the researcher must first grasp how two or more sentences are dependent on each other, which requires processing at many different levels and complicated obstacles. Because of the intricate structure of the Marathi language, this becomes even more difficult. We can see how the discourse anaphora is challenging to resolve in the example below,

"मनिषा मेघनाला भेटायचे म्हणून तिच्याकडे गेली पण ती घरी नव्हती. तिची भेट झाली नाही. तिला तिच्याकडे फार महत्वाचे काम होते."

In this example, "मनिषा" and "मेघना" are antecedents, and here the observable multiple anaphora is "तिच्याकडे", "ती", "तिची" and "तिला". Moreover, the most important thing is that the last two sentences depend on the first initiative sentence.

3.1.4.1 Types of discourse anaphora

There are also some types of discourse anaphora.

a. The definite pronoun anaphora must refer to some noun (phrase) that has already been introduced into the sentence (Houser, M. 2010).

For example,

"आज मला सुरेश भेटला. तो मला खूप विचित्र वाटला."

Today I met Suresh. I found him very strange.

Here, him=is just said the person who met me today is strange.

b. The definite noun phrase anaphora is considered where the noun is called an action-noun and refers to a verb phrase representing an action, process, and happening (Houser, M. 2010).

For example,

"एका माणसाकडे दोन प्रतिभावान सिंह आहेत, ते सिंह शिकारीविषयी चर्चा करत होते."

One man has two talented lions. The lions were discussing the victim.

The lions=the man has two talented lions.

c. One anaphora is an anaphoric noun phrase headed by the word one (Houser, M. 2010).

For example,

"मनीषने त्याच्या वाढदिवशी हिरवा चेंडू मिळाला आणि मला एक निळा."

Manish got a green ball for his birthday, and I got the blue one.

Here, one=ball

d. "Do-it" anaphora is defined as in linguistic the verb phrase do-it is an anaphora; it points to the left toward its antecedent (Webber, 2016).

For example,

"जर राधा नवीन फ्रॉक विकत घेत असेल, तर मी सुद्धा घेईन."

Although Radha buys a new frock, I will do it as well.

Here, do it=buy the new frock.

e. "Do so," anaphora is called the verb phrase anaphoric process, in which the string do-so refers back to an antecedent verb phrase.

Do so anaphora is used as the proof for the internal structure within the verb phrase (Houser, M. J. 2010).

For example,

I ate an apple yesterday in the park, and Riya did so.

Here, do so=ate the apple in the park.

f. Null complements anaphora and behaves like the deep anaphora where the antecedent is missing in the discourse (Cyrino, 2004).

For example,

"मी त्याला जायला सांगितले, पण त्याने नाकारले."

I asked him to leave, but he refused.

Here, 0 or _____=asked him to leave.

Where, (dash "_____") in the sentence, it is null or is an antecedent is missing.

g. Sentential "it" anaphora has investigated sentential anaphora to involve the presence of sentential pro-form, and which does not permit pragmatic control.

For example,

"जरी संदीपच्या पोपटाने त्याचे सफरचंद खाल्ले, परंतु त्याने त्याला त्रास दिला नाही."

Although Sandip's parrot ate an apple of him, it did not bother him.

Here, it=the fact that the parrot ate an apple of Sandip.

h. Sluicing is the process that which the name given the ellipsis process provides that the clause containing it is isomorphic in prior discourse (AnderBois, 2010, August).

For example,

"रसिकाने काहीतरी खाल्ले पण मला माहीत नाही_____"

Rasika ate something, but I do not know _____

Here, _____=something ate

Where, (dash "_____") in the sentence, it is null or a thing that is not known.

i. Gapping is defined as using a gap in zero anaphora in a phrase or clause; it is often referred to as an expression that supplies the necessary information for interpreting the gap (AnderBois, 2010, August).

For example,

"रोशनी लाल रंगाची चप्पल गुरुवारी चालते आणि मधु _____ रविवारी."

Roshni wears red color sandals on Thursday and Madhu _____on Sunday.

Here, (_____) represent=wears dress by Madhu

Where, (dash "_____") in the sentence it is null, or it is the gap in zero anaphora.

j. Stripping (AnderBois, 2010, August).

For example,

"अनिता अर्ध आंबट लोणचे खाते पण तिच्या स्वतःच्या अपार्टमेंटमध्ये नाही."

Anita eats half-sour pickle, but 01, not 02, in her apartment.

Here, 01 represent Anita eats pickel

02 = eat a half-sour pickle

k. "Such" anaphora is a simple and understandable anaphora. (Qiang, Ma, & Wei, 2013)

For example,

"जेंव्हा मयुरीने तिच्याच तोंडात मारली तेंव्हा लहान मुलींमध्ये असे वर्तन घृणास्पद असल्याने तिला शिक्षा झाली."

When Mayuri punched her in the mouth, she was punished for such abominable behavior among little girls.

Here, Aantecedent = Mayuri

Anaphora = her, she used for the Mayuri. Hence it is not confusing to resolve.

3.2 APPROACHES TO ANAPHORA RESOLUTION

While several systems use the same set of parameters, some approaches utilize a classical model that eliminates unlikely options until a few plausible antecedents are obtained. In contrast, other statistical approaches used AI models or techniques to determine the most likely candidate. Whereas computational approaches used to apply some elements may vary, i.e., algorithm, the formula for assigning antecedents, framing rule to find the correct antecedents, rather than a computational issue related to programming language, complexity, etc.

Research on anaphora resolution broadly falls into the following categories, which are as detailed below.

3.2.1 Knowledge-rich approaches

Early research on anaphora resolution typically used a rule-based, algorithmic approach and relied on extensive knowledge. It was based on commonly observed heuristics regarding anaphoric processes. It generally presupposes a complete and adequately parsed input. The evaluation was usually performed manually on a small set of evaluation examples. These approaches can broadly be divided into two categories based on the knowledge employed.

3.2.1.1 Discourse-based approaches

Another traditional method of obtaining the reference of pronouns is discourse-based, called the Centering Theory. This theory models the attention salience of discourse entities and relates it to referential continuity. Centering Theory can be summarized as given below.

Only one entity is the center of attention for each utterance in discourse.

a. The center of utterance is most likely pronominalized (Rule 1 of CT).
b. Consecutive utterances in a discourse maintain the same entity as the center of attention (Rule 2 of CT).

CT-based approaches are attractive from a computational point of view because the information they require can be obtained from the structural properties of utterances alone, as opposed to costly semantic information (Stockyard, 2001).

3.2.1.2 Hybrid approaches

These approaches use several knowledge sources, including syntactic, discourse, morphological, semantic, etc., to rank possible antecedents.

One of the most well-known systems using this approach is that of Lappin and Leass (1994). They use a model that calculates the discourse salience of a candidate based on different factors that are calculated dynamically, and they use this salience measure to rank potential candidates. They do not use costly semantic or real-world knowledge in evaluating antecedents other than gender and number agreement (Lapin & Leas, 1994).

3.2.2 Corpus based approaches

Ge, Hale, and Carina (1998) present a statistical method for resolving pronoun anaphora. They use a small training corpus from the Penn Wall Street Journal Tree-bank marked by co-reference resolution. The base of their method is Hobb's algorithm, but they augment it with a probabilistic model. The kinds of information that are based on their probabilistic model are

1. Distance between the pronoun and its antecedent
2. Syntactic constraints
3. The actual antecedents give information regarding the number, gender, and animaticity
4. Interaction between the head constituent of the pronoun and the antecedent
5. The antecedent's mention count – the more times a referent has occurred in the discourse, the more likely it is to be the antecedent (Ge, Hale, & Carina, 1998)

3.2.3 Knowledge-poor approaches

In recent years, there has been a trend toward knowledge-poor approaches that use machine learning techniques. Soon, Ng and Lim (2001) obtained results comparable to non-learning techniques for the first time. They resolved not just pronouns but all definite descriptions. They used a small annotated corpus to obtain training data to create feature vectors. These training examples were then given to a machine learning algorithm to build a classifier. The learning method is a modification of C4.5 (Quinlan, 1993) called C5, a decision tree-based algorithm. An important point to note about their system is that it is an end-to-end system that includes sentence segmentation, part-of-speech (POS) tagging, morphological processing, noun phrase identification, and semantic class determination (Quinlan, 1993; Soon, Ng, & Lim, 2001).

3.3 CASE STUDY OF ANAPHORA RESOLUTION IN MARATHI TEXT

Marathi is a language from India that belongs to the Indo-Aryan language family. With over 180 million native speakers, it is one of the most widely spoken of India's 22 official languages. Marathi has a diverse morphology and a flexible word order. In Marathi, the default word order is Subject-Object-Verb (SOV). We want to look into linguistic elements that can be utilized to resolve anaphora, mainly using dependency structures as a source of syntactic and semantic information. Our findings revealed that promising performance could be attained by simply utilizing rules-based dependence structure and agreement features.

Marathi pronouns play an essential role in everyday conversation because of their structure. There are different types of Marathi pronouns, which is the most important aspect of the resolution system.

- Personal pronouns (मी, आमही, तुमही): मी तुमहाला माझा ई-मेल देऊ शकते.
- Possessive pronoun (माझा, माझी, तुझा): तुझा पतता मला दे.
- Demonstrative pronoun (तो, ती, ते): ती फार सुंदर दिसते.
- Reflexive pronoun (आपण, आमही, तुमही, तुमहाला): तुमही आमहाला बोलवले होते
- Reciprocal pronoun (एकमेकांचा, एकमेकाला, आपलयाला): आपलयाला गावाला जायचंय.
- Interrogative pronoun (काय, कसे, कोण, का): तू काय करत होतास?

Any language's word order is a major typological characteristic. Without treatment of word order, any study of syntax will be incomplete. Marathi is a verb-final language with a wide range of word order options. [SOV] is the unmarked order. Nevertheless, the locations of distinct words in a Marathi phrase do not always follow a strict sequence.

The first six sentences are variations of the 1–6 sentences in Table 3.1 (Day after tomorrow, Anita will give a book to Sunita). As a result, the placement of some of the chunks in a Marathi sentence can be swapped. In a Marathi sentence, on the other hand, there are several units in which the

Table 3.1 Interchangeable word order in the sentence

S. No.	Marathi sentence	English sentence
1.	अनिता सुनीताला परवा पुस्तक देईल.	Day after tomorrow, Anita
2.	अनिता परवा सुनीताला पुस्तक देईल.	will give a book to Sunita.
3.	अनिता परवा पुस्तक सुनीताला देईल.	
4.	परवा सुनीताला अनिता पुस्तक देईल.	
5.	परवा अनिता सुनीताला पुस्तक देईल.	
6.	सुनीताला परवा अनिता पुस्तक देईल.	

words must appear in a specific order. For example, auxiliary verbs always follow the main verb (e.g., *nighayla hava* निघायला हवं but not *hava nighayla* हवं निघायला), and case suffixes and postpositions always follow the nouns (e.g., *Manisha ne* मनीषाने but not *ne Manisha* ने मनीषा). Adjectives precede the noun they modify (e.g., *laal saadi* लाल साडी but not *saadi laal* साडी लाल).

Because of this unspecified order of the words in the sentence, we need to be part of a speech tagger to determine the correct tag for the word. This is the first step in the resolution of an anaphora. Hence, we developed a rule-based POS tagger for Marathi text with the help of the N-gram.

3.3.1 Development of POS tagger

Tagging Marathi text with POS tags is a challenging task as Marathi is an inflective language where suffixes are added to words to add semantics to the context. The major challenge in POS tagging is to address this ambiguity in the possible POS tags for a word. We have developed a POS tagger that will assign POS to the word in a sentence provided as input to the POS tagger system with the help of a rule-based approach. The work of the Marathi POS tagging is based on different types of rules. The rules of POS tagging for Marathi are based on the categories of words it belongs to. Marathi is a Prakrit language, meaning it is derived from Sanskrit. Marathi is a verb-final language and a free-ordered language. The goal is to create a POS tagger for Marathi.

One of the most important aspects of NLP is using a combination of taggers. The relevance of the combined tagger is that when one tagger fails to tag a word, it is sent to another, known as sequential backoff tagging. Kalpana and Namrata discussed that the default tagger was chosen as the backoff tagger because, during training, if it failed to tag the word, the unigram tagger would be passed to the default tagger to tag it with the default value of noun (NN).

The unigram tagger is a subclass of the n-gram tagger. N-gram has two subclasses: they are most widely used, i.e., bigram and trigram, respectively. As the name implies, the unigram tagger assigns the part of the speech tag to only one word in its context. It is a context-based tagger with a single-word context. As the name implies, the bigram tagger tags two words: one is the previous tag, and the other is the current tagged word. The result of bigram tagging is not based on the word's context, whereas the unigram tagger does not care about the preceding word and only tags the current one. As a result of combining both taggers, the result is superior. In general, we first spilt or tokenized the sentences into words, and then applied the tagger combination.

Algorithm for POS tagger:

Step 1: Take Marathi text as input
Step 2: Create a list of tokenized words from sentences of input text
Step 3: Apply a rule-based POS tagger to the tokenized text
Step 4: If the tag is not correct, repeat Step 3
Step 5: Stop

Figure 3.1 depticts the workflow of the POS tagger with N-gram methods, and the issues found with Marathi text have also been discussed.

Main issues encountered while designing a rule-based POS tagger

1. The segmentation ambiguity is maintained by using a full stop at the end of the segmentation, such as (व.पु.काळे). In English, a full stop signifies the end of a sentence.
2. Another issue is that the term "नवी-दिलली" in Marathi is a proper noun, yet it is tokenized as two separate tokens, "नवी" and "दिलली".
3. Some complications arise while tagging Marathi language material since it has words such as (बोलता-बोलता, दुःख:) which include the hyphen

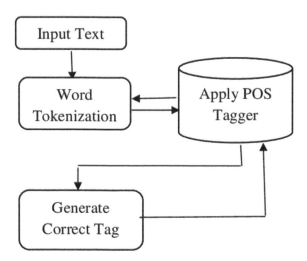

Figure 3.1 Workflow of the POS tagger.

(-) and colon (:) but no other punctuation mark. However, when a hyphen appears in a sentence, it is considered a separate token. In addition, "बोलता" and "बोलता" take into account two different tokens in the statement.

4. Issue related to encoding:

The Marathi content is written in a variety of fonts. The document is unintelligible or unusable when a language character is not shown correctly on a computer. When a document is created on one computer with a given operating system, it may not appear on another machine with a different configuration. Some characters are accessible in these circumstances, but some are not entirely displayed, and there may be hollow circles or symbolic interpretations. For instance, it is printed as. For example, केंद्र is printed as केंद्.

5. As a result, all font types must be traced and converted to a single format, which is time-consuming. Two verbs are adjacent to each other when tagging the sentence. The first verb serves as the main verb, indicating an incomplete action, while the second, indicating a complete action, serves as an auxiliary verb. However, in specific sentences, there were multiple verbs. For example, ममता पवनला पाणी नेऊन देत होती." in this case, the verbs "नेऊन" and "देत" are contiguous, and there is only one auxiliary verb, which is "होती".

Kalpana and Namrata have demonstrated the use of a combined tagger to train their data, where they trained 17,191 words of Marathi and achieved 30% better results than statistical and syntactically based tagging.

3.3.2 Anaphora resolution system architecture

Figure 3.2 explains the architecture of the Marathi anaphora resolution system and discusses about how we apply the rule-based approach for resolving the anaphora and what issues were found during the resolution process. A detailed discussion of the architecture is mentioned below.

3.3.2.1 Pronominal anaphora resolution using rule-based approach

3.3.2.1.1 Simple sentence

This rule is framed to resolve an anaphora in a simple sentence that contains one antecedent and only one anaphora. This type of sentence is not as difficult to resolve as others.

  i. If PRP ("तो"," ती"," ते") present in the second sentence
 ii. Then pronoun refers to the proper antecedent
iii. Else pronoun is not found properly

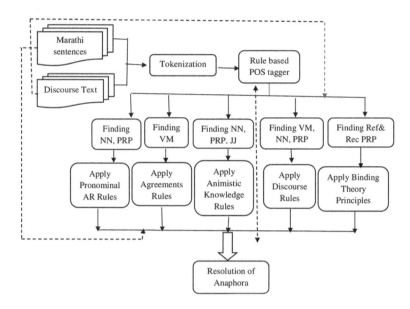

Figure 3.2 The architecture of the Marathi Anaphora resolution system.

3.3.2.1.2 Complex sentence

These rules are basically based on complex sentences with more than one antecedent and only one anaphora present; hence, these types of sentences are quite difficult to resolve.

   i. If PRP ("तो", "ती", "ते") present in the second sentence
   ii. Then match the successive word of a proper noun (NNP) is a noun (NN) and the successive word of a pronoun (PRP) is adjective (JJ)
   iii. Then an anaphora refers to the noun (NN)
   iv. Else anaphora is not resolved properly

3.3.2.2 Anaphora resolution using gender and number agreement

3.3.2.2.1 Gender agreement

This agreement in the anaphora resolution in the text of Marathi is according to the gender of the person, whether male or female, or neutral. For example,

   1. नितीन मिठाईच्या दुकानात गेला व तो रसमलाई खात होता.
   2. आश्विनी मिठाईच्या दुकानात गेली व ती रसमलाई खात होती.

The examples above show two types of verbs: compound verbs and auxiliary verbs, "गेला" and "होता". At the end of the word, the verb is "kana" ("ा"). So it is a male-dominated genre. Similarly, the verbs in the second sentence use the compound verb "गेली" and the auxiliary verb "होती", while the velanti "ी" identifies the person's gender. We can quickly determine which anaphora refers to which antecedent with this characteristic.

Rules for gender agreement:

If a compound or auxiliary verb ends with "ा"
　　Then noun (NNP) and pronoun (PRP) refer to masculine.
If compound or auxiliary verb ends with velanti "ी"
　　Then noun (NNP) and pronoun (PRP) refer to as feminine.

With examples 1 and 2 above, we can easily understand the rules for the gender agreement.

Major issues faced during resolution:

While resolving the anaphora of two antecedents who have the same gender, it was somewhat difficult to solve the anaphora. For example,

"रमेश व गणेश जिवलग मित्र आहेत, तो त्याची काळजी खूप घेतो."

In the above example, same-gender antecedents are present that are "**रमेश**" and "**गणेश**" and two anaphora like "**तो**" and "**त्याची**" we are not able to identify which proper noun is being referred to.

3.3.2.2.2 Number agreement

The number agreement is based on POS tagging of words, and we can determine if a word is a singular or plural using a POS tagger. For example,

1. प्रथमेश आणि आकाश मित्र आहेत ते दोघे सोबत सहलीला जाणार होते.
2. मृणाली स्पर्धा जिंकली कारण ती हुशार होती. or राजेश स्पर्धा जिंकला कारणतो हुशार होता.

Example 1 above shows the anaphora "**ते**" pointing toward "**परथमेश**" and "**आकाश**" which is plural because the verb ends with the "matra" ("े"). Similarly, in example 2, the anaphora "**ती**" or "**तो**" is referred to for the singular antecedent, which is, respectively, "**मृणाली**" or "**राजेश**".

The rule for number agreement:

If the verb ends with matra "े"
　　Then a noun or pronoun refers to plural
If the verb ends with kana "ा" and velanti "ी"
　　Then a noun or pronoun refers to singular.

Major issues faced during resolution:

When a verb ends in the matra "ं", the noun and the pronoun are plurals, but when we talk about revered people like Abdul Kalam, Shivaji Maharaj, etc., in Marathi, the pronoun is "ते". It is addressed to a specific individual to show respect for that individual. This problem was addressed when the anaphora was resolved based on the number agreement. For example,

"अब्दुल कलाम राष्ट्रपती होते आणि ते खूप प्रसिद्ध संशोधकही होते."

In the above example, the "अब्दुल कलाम" is only one antecedent in the sentence, and to give him respect, we use as the pronoun or anaphora is "ते". Moreover, therefore, the verb ended with "ते."

3.3.2.3 Anaphora resolution using the animistic knowledge

Animistic knowledge is based on those who represent living and inanimate beings. In the below example,

मनीषने समोसा खाल्ला, तो तिखट होता.

Here, the above example shows "मनीष" as a proper noun (NNP) and "समोसा" as the noun (NN). On the other hand, "तो" is a pa pronoun (PRP), and the successive word "तिखट" is an adjective (JJ). According to the rule, the pronoun "तो" is referred to as the noun "समोसा".

> If match the successive words of a proper noun (NNP) is the noun (NN) and the pronoun (PRP) is the adjective (JJ).
> Then the pronoun refers to the noun (NN).

3.3.2.4 Discourse anaphora resolved using rule-based approach

3.3.2.4.1 Possessive pronoun

This rule applies to possessive pronouns in Marathi such as "त्यांनी", "त्यांचे", "त्यांची", etc. We found in our database that it contains the possessive pronoun. Another thing we are trying to resolve in an anaphora is that various sentences are linked together and dependent on one another. Assume three sentences are linked together, and one of the statements is cut. The sentences are no longer intelligible. The issue of who the subject is and what work is done for the discourse is difficult.

The rule for resolving an anaphora:

If the discourse consists of three sentences:

Then we found that other sentences were linked together:

> If (.) full stop is present after the VM:
> Then the first word of the second and third sentences is PRP
> If PRP is found in the second and third sentences
> Then go back to the first sentence and check for the noun or
> a proper noun, and the antecedent is found
> Else: Pronoun is not found

Anaphora resolved properly in discourse
For example,
Input text: गाडगेबाबांच्या कीर्तनाला दहा-दहा हजार लोक जमू लागले. त्यांच्या पायावर डोकी ठेवायला लोकांची झिम्मड उडू लागली. ते कुणाला पाया पडू देत नसत.
Anaphora resolved: गाडगेबाबांच्या कीर्तनाला दहा-दहा हजार लोक जमू लागले. गाडगेबाबांच्या पायावर डोकी ठेवायला लोकांची झिम्मड उडू लागली. गाडगेबाबा कुणाला पाया पडू देत नसत.

3.3.2.4.2 First-person possessive and third-person plural possessive pronoun

The resolution of the anaphora makes it rather difficult to grasp the correct anaphora for the specific antecedent because the person plural pronouns are present in the sentences combined. Which antecedent is correct for the anaphora is a source of consternation for the system.
The rule for resolving an anaphora:
If the discourse is in two or three sentences:
Then we found that other sentences were linked together

> If PRP is between sentence and (.) full stop present after the VM:
> Then the first word of the second sentence is PRP
> If "त्यांचा/त्यांची/त्यांचे" is found in the second sentence:
> Then third-person possessive pronoun refers to the antecedent
> NN or NNP
> Else: PRP is a first-person possessive pronoun
> Else: Pronoun is not found

Anaphora resolved properly as a possessive pronoun
Input text: भाऊरावांना भेटावे म्हणून त्यांच्याकडे गेलो. त्यांची भेट झाली. मला खूप समाधान वाटले.
Anaphora resolved: भाऊरावांना भेटावे म्हणून भाऊरावांकडे गेलो. भाऊरावांची भेट झाली. मला खूप समाधान वाटले.
Issue: (Object Pronoun, Possessive Pronoun, Reflexive Pronoun)
We found a discourse in our database where three or more pronouns are present in a single discourse. This form of discourse is extremely difficult to resolve since the system does not accurately identify the anaphora for the antecedent. As a result, these sentences do not resolve appropriately.

For example,

Input text: अण्णा *त्याच्याजवळ* गेले.*त्यांनी त्याला* आपल्या दोन्ही हातांनी अलगत उचलले. *स्वतःच्या* बिछान्यावर आणून झोपवले.*स्वतःची* कांबळी *त्याच्याअंगावर* घातली.*त्याला* अगदी आपल्या पोटाशी धरले. *त्या* उबेत *तो* मुलगा गाढ झोपी गेला.

Here, in this example,

- "अण्णा" is the antecedent or the proper noun tagged as NNP.
- "त्याच्याजवळ", "त्याच्या", "त्याला"," तो" this anaphora's is referred for the unknown antecedent. These are the object pronouns.
- "त्यांनी" is used as anaphora, which is referred to as the antecedent NNP. Moreover, it is the possessive pronoun.
- "स्वतःच्या" and "स्वतःची" are also used for NNP. Moreover, these are reflexive pronouns.

To overcome this above-mentioned issue, we have concentrated on the Binding theory by Chomsky.

3.3.2.5 Anaphora resolution using Binding theory

Chomsky initially stated three syntactic principles that govern the distribution of DPs in a phrase as a module of Government and Binding Theory (Chomsky, 1956), where he stated three syntactic principles that govern the distribution of DPs in a sentence. Theoretically, retrieving the semantic content of a pronoun appears to be a problem involving meaning, i.e., human language semantics; however, the distribution of pronouns in a sentence and the relationships between their denotations appear to be constrained by structural, or syntactic, constraints. The debate over the syntax/semantics interface in human languages is centered on Binding theory.

The Binding theory is concerned with defining the conditions of reference dependency between expressions in a given language. In other words, the theory explains how noun phrases and their grammatical antecedents might be related.

The following sentence, for example, is grammatically correct insofar as the pronoun and proper name do not refer to the same person. However, it is ungrammatical if "तीला" is referred to "मीना".

"तिला वाटते मीना चांगली मुलगी आहे."

However, if the name and pronoun are changed, then the pronoun may always be understood,

but it can also refer to मीना.

"मीनाला वाटते ती चांगली मुलगी आहे."

The Binding theory is a crucial topic of study in inquiries into the alliance between syntax and semantics because of this connection between distribution and interpretation. Theories and formalisms focus on discourse, such as the depiction of discourse.

We evaluated the reflexive and reciprocal pronouns to resolve the Marathi text. While researching for the database, we discovered many reflexive and reciprocal pronouns in the conversation, necessitating the use of a binder to resolve anaphora. Swataah (स्वतः) performs the function of a reflexive pronoun. There is also the long-distance reflexive aapan (आपण), which has to be locally free within NPs and serves as the object of prepositions that assign their own.

3.3.1.3.1 *Principles of Binding theory*

Based on the notion of binding, conventional Binding theory analyses configurations in which two DPs (determiner phrases) appear in a sentence and can, should, or should not be co-referential. As a result, the heart of the Binding theory is made up of three principles that describe various ways in which a DP can, must, or must not be bound within a phrase.

a. **Principle A**
   Table 3.2 shows the local domain to which a reflexive pronoun must be bound. It signifies that the antecedents and anaphora in the sentences have a co-reference to one another – the reflexive pronoun "स्वतःचा" bound with its antecedent "मनोज" in Table 3.2.

b. **Principle B**
   Table 3.3 shows the limited domain in which a non-reflexive pronoun must not be bound. In Table 3.2, the antecedent "मंगेश" is not connected with the non-reflexive pronoun "त्याचा," as seen in the table below. The reflexive pronoun is connected with different antecedents "श्याम" in Table 3.3.

c. **Principle C**
   Table 3.4 shows a referential expression must not be bound. Where the antecedents and the anaphora are not bound.

Here, numbers 1 and 2 represent the indexing of the antecedents to better understand the principles of Binding theory.

Table 3.2 Example of principle A

S. No.	English sentence	Marathi sentence
1.	*Robert1 hates himself1	मंगेशला1 स्वतःचा1 रागआला.
2.	Robert1 thinks that Martin2 hates himself2	मंगेशला1 वाटतेकी श्यामला2 स्वतःचा2 रागआला.

* indicates the reflexive pronoun bounding within its local domain

Table 3.3 Example of principle B

S. No.	English sentence	Marathi sentence
1.	*Robert1 hates him1	मंगेशला1 स्वतःचा1 रागआला
2.	Robert1 hates him2	मंगेशला1 त्याचा2 रागआला.
3.	Robert1 thinks that Martin2 hates him2	मंगेशला1 वाटते की श्यामला2 स्वतःचा2 रागआला.

* indicates the reflexive pronoun bounding within its local domain

Table 3.4 Example of principle C

S. No.	English sentence	Marathi sentence
1.	*Robert1 hates Robert1	मंगेशला1 स्वतःचा1 रागआला.
2.	Robert1 hates him2	मंगेशला1 त्याचा2 रागआला.
3.	*He1 thinks that Martin2 hates Robert1	त्याला1 वाटते श्यामला2 मंगेशचा1 रागआला.

* indicates the reflexive pronoun bounding within its local domain

3.4 CASE STUDY OF ANAPHORA RESOLUTION IN MARATHI WITH PYTHON

This study is based on the rule-based approach, and we have resolved an anaphora with the NLTK toolkit of the Python programming language.

3.4.1 Database

As mentioned in Section 3.1, the corpus of the Marathi discourse has not been available; hence, we first created the corpus. This study (Kalpana and Namrata) collected the database, including antecedents and an anaphora. From the Marathi Balbharati Textbook, we have chosen the discourse sentences of the first- to eighth-grade chapters. We have chosen a total of 1,000 discourses. Some are simple sentences, some are complex, and many discourses conflict to resolve an anaphora. Table 3.5 shows the sample of databases collected during the study.

3.4.2 Preprocessing

After collecting the database, the next stage is preprocessing, which includes tokenizing the discourse into separate sentences and then tokenizing the sentences into words. Figure 3.3 shows that the outcome of the preprocessing stage of the anaphora resolution is tokenization.

Table 3.5 Sample database

Simple	Complex	Discourse (Conflict)
शशी बासरी वाजवतो, तो चांगला बासरी वादक आहे.	रश्मीने राणीला बाजारात नेलं आणि ती खूप थकली	म्हातारा त्या तरुणाला म्हणाला, "राजा, तुझं इंग्लंडला जाणं पक्कं झालं ना?" "असं काय काका मला विचारता? मी पासपोर्ट मिळवला, व्हिसा काढला, नवीन कपडे टेलरने परवाच शिवून दिले. चार सूट आणि इतर नेहमीचे कपडे घेतले आहेत. तुम्ही दिलेले घड्याळ, तर मी आताच वापरायला काढलं आहे."
सोहमचे लग्न आहे, तो खूप खुश आहे.	राजेशने रवीला फोन केला पण तो तिथे नव्हता	"मनिषा मेघनाला भेटायचे म्हणून तिच्याकडे गेली पण ती घरी नव्हती. तिची भेट झाली नाही. तिला तिच्याकडे फार महत्वाचे काम होते."
सुरेशने गणित सोडवले, तो फार हुशार आहे.	मधुराने मालतीला खूप सतावले ती तिची मुलगी आहे.	पल्यांनी बनवलेल्या खोलीत अण्णा आणि मनोज बसले होते. ते त्याला गोष्ट सांगत होते. तोही त्यांची गोष्ट मन लाऊन ऐकत होता.
राजू रोज दोन मैल चालतो, तो खूप जाड आहे.	मधुर विराजची मदत करत होता कारण तो खूप चांगला मुलगा आहे.	रामाने रावणाला मारले. ही गोष्ट फार पुराणिक आहे पण यात ज्याने मारले तो देव होता आणि जो मेला तो राक्षस. तो त्याच्या कर्माने मारला गेला.
शामाने थट्टा केली, ती खूप विनोदी आहे.	स्वरा मिराशी भांडली कारण ती फार भांडकुदळ आहे.	कविताने आणलेली बाहुलीला सुधाला नको होती. कारण तिच्याकडे आधीच खूप बाहुल्या होत्या आणि तिला नवीन टेडीबिअर हवा होता. पण तिला तो मिळाला नाही म्हणून ती रुसली होती.

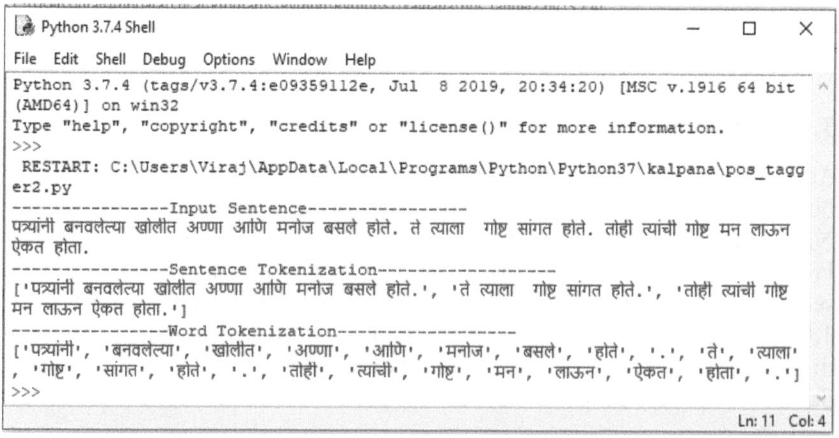

```
Python 3.7.4 Shell                                          —    □    ×
File  Edit  Shell  Debug  Options  Window  Help
Python 3.7.4 (tags/v3.7.4:e09359112e, Jul  8 2019, 20:34:20) [MSC v.1916 64 bit
(AMD64)] on win32
Type "help", "copyright", "credits" or "license()" for more information.
>>>
 RESTART: C:\Users\Viraj\AppData\Local\Programs\Python\Python37\kalpana\pos_tagg
er2.py
-----------------Input Sentence-----------------
पत्र्यांनी बनवलेल्या खोलीत अण्णा आणि मनोज बसले होते. ते त्याला  गोष्ट सांगत होते. तोही त्यांची गोष्ट मन लाऊन
ऐकत होता.
-----------------Sentence Tokenization-------------------
['पत्र्यांनी बनवलेल्या खोलीत अण्णा आणि मनोज बसले होते.', 'ते त्याला  गोष्ट सांगत होते.', 'तोही त्यांची गोष्ट
मन लाऊन ऐकत होता.']
-----------------Word Tokenization-----------------
['पत्र्यांनी', 'बनवलेल्या', 'खोलीत', 'अण्णा', 'आणि', 'मनोज', 'बसले', 'होते', '.', 'ते', 'त्याला'
, 'गोष्ट', 'सांगत', 'होते', '.', 'तोही', 'त्यांची', 'गोष्ट', 'मन', 'लाऊन', 'ऐकत', 'होता', '.']
>>>
                                                              Ln: 11  Col: 4
```

Figure 3.3 Preprocessing of text.

3.4.3 Post-processing

After preprocessing, moving toward post-processing where we have tagged the words with the rule-based combined tagger mentioned in Section 3.1, we have concentrated on the Marathi POS tagger in this stage. With the help of the NLTK toolkit and different types of libraries (for example, pickle, dump, etc.), we have trained and tested the words to produce the correct tag of the word in the Marathi text. Here, we have tagged with a rule-based approach for a total of 17,191 words. After that, using these tagged words, we have to find antecedents and an anaphora pair from the text and try to detect the correct anaphora to the correct antecedents with the help of the above mentioned in Sections 3.2–3.4.

Finally, the results are calculated by the following formula, and Figure 3.4 shows the result of the POS tagger for Marathi text.

$$\text{Accuracy} = \text{Number of Anaphora Resolved} / \\ \text{Total Number of Anaphora Present in the Text}$$

Figure 3.5 depticts the final resolution of anaphora for Marathi text, and the relevant pair of antecedents and anaphora have been found in the given text.

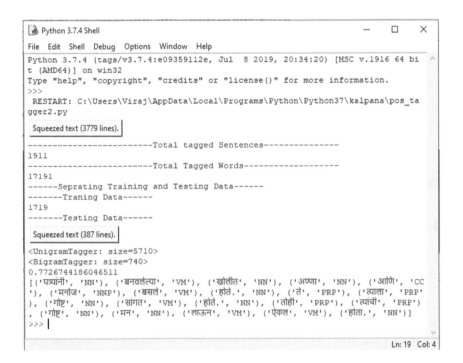

Figure 3.4 POS tagging of text.

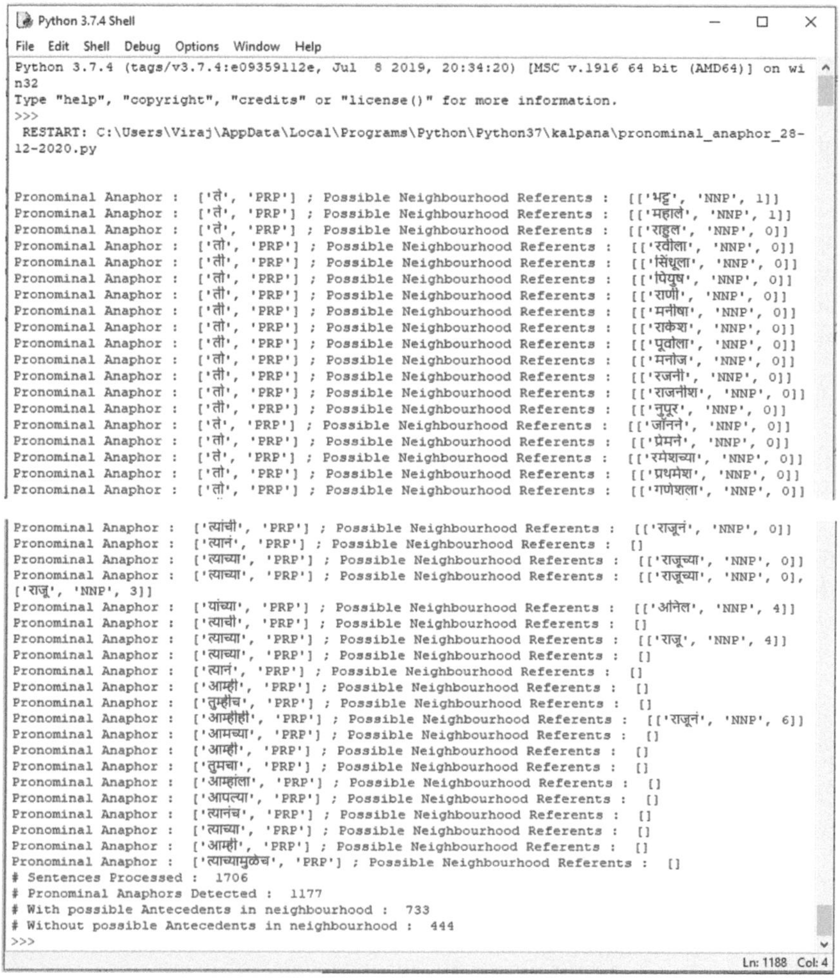

```
Python 3.7.4 Shell                                          —   □   ×
File  Edit  Shell  Debug  Options  Window  Help
Python 3.7.4 (tags/v3.7.4:e09359112e, Jul  8 2019, 20:34:20) [MSC v.1916 64 bit (AMD64)] on wi
n32
Type "help", "copyright", "credits" or "license()" for more information.
>>>
 RESTART: C:\Users\Viraj\AppData\Local\Programs\Python\Python37\kalpana\pronominal_anaphor_28-
12-2020.py

Pronominal Anaphor :   ['ते', 'PRP'] ; Possible Neighbourhood Referents :   [['भट्टू', 'NNP', 1]]
Pronominal Anaphor :   ['ते', 'PRP'] ; Possible Neighbourhood Referents :   [['महालें', 'NNP', 1]]
Pronominal Anaphor :   ['ते', 'PRP'] ; Possible Neighbourhood Referents :   [['राहुल', 'NNP', 0]]
Pronominal Anaphor :   ['ती', 'PRP'] ; Possible Neighbourhood Referents :   [['रवीला', 'NNP', 0]]
Pronominal Anaphor :   ['ती', 'PRP'] ; Possible Neighbourhood Referents :   [['सिंधूला', 'NNP', 0]]
Pronominal Anaphor :   ['ती', 'PRP'] ; Possible Neighbourhood Referents :   [['पियूष', 'NNP', 0]]
Pronominal Anaphor :   ['ती', 'PRP'] ; Possible Neighbourhood Referents :   [['राणी', 'NNP', 0]]
Pronominal Anaphor :   ['ती', 'PRP'] ; Possible Neighbourhood Referents :   [['मनीषा', 'NNP', 0]]
Pronominal Anaphor :   ['ती', 'PRP'] ; Possible Neighbourhood Referents :   [['राकेश', 'NNP', 0]]
Pronominal Anaphor :   ['ती', 'PRP'] ; Possible Neighbourhood Referents :   [['पूर्वाला', 'NNP', 0]]
Pronominal Anaphor :   ['ती', 'PRP'] ; Possible Neighbourhood Referents :   [['मनोज', 'NNP', 0]]
Pronominal Anaphor :   ['ती', 'PRP'] ; Possible Neighbourhood Referents :   [['रजनी', 'NNP', 0]]
Pronominal Anaphor :   ['ती', 'PRP'] ; Possible Neighbourhood Referents :   [['राजनीश', 'NNP', 0]]
Pronominal Anaphor :   ['ती', 'PRP'] ; Possible Neighbourhood Referents :   [['नूपूर', 'NNP', 0]]
Pronominal Anaphor :   ['ते', 'PRP'] ; Possible Neighbourhood Referents :   [['जॉनिन', 'NNP', 0]]
Pronominal Anaphor :   ['ती', 'PRP'] ; Possible Neighbourhood Referents :   [['प्रेमन', 'NNP', 0]]
Pronominal Anaphor :   ['ते', 'PRP'] ; Possible Neighbourhood Referents :   [['रमेशच्या', 'NNP', 0]]
Pronominal Anaphor :   ['ती', 'PRP'] ; Possible Neighbourhood Referents :   [['प्रध्मेश', 'NNP', 0]]
Pronominal Anaphor :   ['तो', 'PRP'] ; Possible Neighbourhood Referents :   [['गणेशला', 'NNP', 0]]


Pronominal Anaphor :   ['त्यांची', 'PRP'] ; Possible Neighbourhood Referents :   [['राजूनं', 'NNP', 0]]
Pronominal Anaphor :   ['त्यानं', 'PRP'] ; Possible Neighbourhood Referents :   []
Pronominal Anaphor :   ['त्याच्या', 'PRP'] ; Possible Neighbourhood Referents :   [['राजूच्या', 'NNP', 0]]
Pronominal Anaphor :   ['त्याच्या', 'PRP'] ; Possible Neighbourhood Referents :   [['राजूच्या', 'NNP', 0],
['राजू', 'NNP', 3]]
Pronominal Anaphor :   ['यांची', 'PRP'] ; Possible Neighbourhood Referents :   [['अनिल', 'NNP', 4]]
Pronominal Anaphor :   ['त्याची', 'PRP'] ; Possible Neighbourhood Referents :   []
Pronominal Anaphor :   ['त्याच्या', 'PRP'] ; Possible Neighbourhood Referents :   [['राजू', 'NNP', 4]]
Pronominal Anaphor :   ['त्याच्या', 'PRP'] ; Possible Neighbourhood Referents :   []
Pronominal Anaphor :   ['त्यानं', 'PRP'] ; Possible Neighbourhood Referents :   []
Pronominal Anaphor :   ['आम्ही', 'PRP'] ; Possible Neighbourhood Referents :   []
Pronominal Anaphor :   ['तुम्हीच', 'PRP'] ; Possible Neighbourhood Referents :   []
Pronominal Anaphor :   ['आम्हीही', 'PRP'] ; Possible Neighbourhood Referents :   [['राजूनं', 'NNP', 6]]
Pronominal Anaphor :   ['आमच्या', 'PRP'] ; Possible Neighbourhood Referents :   []
Pronominal Anaphor :   ['आम्ही', 'PRP'] ; Possible Neighbourhood Referents :   []
Pronominal Anaphor :   ['तुमचा', 'PRP'] ; Possible Neighbourhood Referents :   []
Pronominal Anaphor :   ['आम्हाला', 'PRP'] ; Possible Neighbourhood Referents :   []
Pronominal Anaphor :   ['आपल्या', 'PRP'] ; Possible Neighbourhood Referents :   []
Pronominal Anaphor :   ['त्यानंच', 'PRP'] ; Possible Neighbourhood Referents :   []
Pronominal Anaphor :   ['त्याला', 'PRP'] ; Possible Neighbourhood Referents :   []
Pronominal Anaphor :   ['आम्ही', 'PRP'] ; Possible Neighbourhood Referents :   []
Pronominal Anaphor :   ['त्याच्यामुळेच', 'PRP'] ; Possible Neighbourhood Referents :   []
# Sentences Processed :  1706
# Pronominal Anaphors Detected :  1177
# With possible Antecedents in neighbourhood :   733
# Without possible Antecedents in neighbourhood :   444
>>>
                                                                    Ln: 1188  Col: 4
```

Figure 3.5 Detected pronominal and discourse antecedents and anaphora.

3.5 CONCLUSION

The Marathi anaphora resolution is the inspiration for this piece. There has not been much progress, as there has not been in Marathi. Because Marathi is a free-order language, we have had to deal with several challenges in terms of linguistic structure. One of the researcher's most challenging tasks is resolving the anaphora. The researcher has frequently resolved the pronominal anaphora. We have been attempting to resolve the anaphora for the Marathi text because many efforts on the anaphora have been made for various Indian languages but not for the Marathi text. Because anaphora varies from language to language, we tried to resolve the anaphora for the Marathi text. We concentrated on the agreement on gender and number, as well as animist knowledge, to resolve the anaphora. Also, concentrate on the Binding theory, which is the key area of the resolution of discourses. To resolve the Marathi text, we used the database as the Marathi Balbharati Textbook, first- to eighth-grade chapters.

Compared to other approaches for anaphora resolution, our rules-based approach has proven to be more effective in resolving anaphora in Marathi text. The performance of the anaphora resolution system for Marathi text and an anaphora pair from the sentence have been detected as antecedents. Moreover, we have to try to resolve all the issues tackled during the study in the future.

REFERENCES

AnderBois, S. (2010, August). Sluicing as anaphora to issues. In *Semantics and linguistic theory* (Vol. 20, pp. 451–470). Linguistic Society of America. https://doi.org/10.3765/salt.v20i0.2574

Chomsky, N. (1956). Three models for the description of language, *IRI Transactions on Information Theory*, 2(3), 113–124.

Cyrino, S. (2004). Null Complement Anaphora and Null Objects in Brazilian Portuguese, trabalhoapresentado no Workshop on Morphosyntax, Universidad de Buenos Aires. submetido a MIT Working Papers.

Gee, N., Hale, J., & Carina, E. (1998, August). A statistical approach to anaphora resolution. In *Proceedings of the sixth workshop on very large corpora* (Vol. 71, p. 76). Sixth Workshop on Very Large Corpora. Aclanthology. https://aclanthology.org/W98-1119

Houser, M. J. (2010). *On the anaphoric status of do so*. https://linguistics.berkeley.edu/~ mhouser/Papers/do_so_status.pdf.

Houser, M. (2010). *The syntax and semantics of do so anaphora, Unpublished PhD dissertation*, University of California at Berkeley.

https://cart.ebalbharati.in/BalBooks/ebook.aspx. Marathi Balbharati Textbook.

Lappin, S., & Leass, H. J. (1994). An algorithm for pronominal anaphora resolution. *Computational Linguistics*, 20(4), 535–561.

Qiang, Y., Ma, R., & Wei, W. (2013). Discourse anaphora resolution strategy based on syntactic and semantic analysis. *Information Technology Journal*, 12(24), 8204–8211.

Quinlan, J. R. (1993). *C4.5: Programs for machine learning*. Morgan Kaufman: San Mateo, CA.

Soon, W. M., Ng, H. T., & Lim, D. C. Y. (2001). A machine learning approach to co-reference resolution of noun phrases. *Computational Linguistics*, 27(4), 521–544.

Stuckardt, R. (2001). Design and enhanced evaluation of a robust anaphor resolution algorithm. *Computational Linguistics*, 27(4), 479–506.

Webber, B. L. (Ed.). (2016). *A formal approach to discourse anaphora*. Routledge.

Chapter 4

A review of the approaches to neural machine translation

Preetpal Kaur Buttar and Manoj Kumar Sachan
Sant Longowal Institute of Engineering and Technology

4.1 INTRODUCTION

The ability to effortlessly translate a piece of text written in one language into another has remained a centuries-old desire of human civilization. Before the emergence of computer technologies, this task was the sole responsibility of human translators. However, after the advent of computers, especially the World Wide Web (www), the world has come closer. People from different cultures speaking different languages are generating and consuming the information/content in their native languages on www. In order to interact with people speaking other languages, machine-assisted translation has become a dire need. Translating a piece of text from one natural language to another using a computer is called *machine translation*. In other words, machine translation means teaching a machine to learn to translate automatically. Machine translation has become a crucial research field due to increased globalization and its social, government, military, and business requirements. Recently, the introduction of deep learning techniques has revolutionized research in the fields of image processing (Krizhevsky et al., 2012), speech recognition (Graves et al., 2013), and natural language processing (Cho et al., 2014b), as they can learn complex tasks from the data automatically. In recent years, deep learning in machine translation has achieved significant hype, which gave birth to a new approach to machine translation known as NMT (Cho et al., 2014a; Kalchbrenner & Blunsom, 2013b; Sutskever et al., 2014).

In this literature review, we briefly discussed the different approaches to machine translation in Section 4.2. Section 4.3 formulates and explains the NMT task. Section 4.4 discusses the encoder-decoder model indepth, while Section 4.5 emphasizes recurrent neural networks (RNNs) as encoder-decoder models. Section 4.6 explains the long short-term memory (LSTM) cells; Section 4.7 focuses on the attention mechanism in NMT; Section 4.8 discusses the recent advances in NMT, including convolutional neural network (CNN)-based NMT and the transformer models; and Sections 4.9 and 4.10 are attributed to the two significant challenges of low-resource NMT and the vocabulary coverage problem, respectively. Lastly, Section 4.11 concludes the review with pointers to future research directions.

 DOI: 10.1201/9781003244332-4

4.2 MACHINE TRANSLATION APPROACHES

Research in the field of machine translation began in the 1950s. The *First International Conference on Machine Translation* was organized in 1952 (Hutchins, 1997). Three primary approaches that emerged in the field of machine translation are rule-based machine translation (RBMT) (Arnold et al., 1993), statistical machine translation (SMT) (Brown et al., 1990), and NMT (Cho et al., 2014a).

In RBMT, a source sentence is translated into the target language through lexical transfer and a large rule base to capture the grammatical properties of the source and target languages (Arnold et al., 1993). Words are directly translated using a bilingual dictionary. The sentences may require local reordering to achieve grammatically correct translations on the target side. The major problem with this approach is that natural languages are inherently complex. Therefore, it is very time-consuming and challenging to incorporate the behavior of these languages into an automated system.

The era of SMT began in the 1990s with the seminal work of Brown et al. (1990). SMT approaches dominated the field of machine translation until the arrival of deep learning approaches in 2015. SMT approaches employ bilingual parallel corpora to generate a probability-based translation. SMT approaches can be categorized into word-based approaches, syntax-based approaches, phrase-based approaches, and hierarchical phrase-based approaches (Koehn, 2010). Before NMT, phrase-based SMT (PBSMT) was the dominant machine translation approach, which performed the translation based on phrases. Phases ensure the inclusion of context while performing the translation. The various components of PBSMT are: pre-processing, sentence alignment, word alignment, phrase extraction, phrase feature preparation, and language model training (Koehn et al., 2003). SMT approaches are automated compared with rule-based approaches, where one had to handcraft the rule base. Among the drawbacks of this approach, the major issue is that the different components, i.e., translation models, language models, reordering models, etc., need to be tuned separately. Separately tuned components are challenging to integrate (Luong et al., 2015a). Many translation decisions are determined locally, phrase-by-phrase, and long-distance dependencies are often ignored. This resulted in the degradation of translation performance.

Due to more data, more computing power, and access to advanced neural networks and algorithms, the application of deep neural architectures for various learning tasks has seen tremendous growth in recent years. With the proliferation of deep learning in 2010, more and more complex learning tasks which were traditionally considered difficult witnessed improved performance (Krizhevsky et al., 2012; Maas et al., 2013). Due to their exceptional performance, deep learning also found potential applications in many NLP tasks, such as speech recognition (Chorowski et al., 2014) and named entity recognition (Collobert et al., 2011). Kalchbrenner and Blunsom (2013b) first proposed the use of deep neural networks for machine translation. NMT approaches outperformed SMT for 75% of the shared tasks in WMT 2016 (Bojar et al., 2016).

4.3 FORMULATION OF THE NMT TASK

NMT is a sequence-to-sequence modeling task, i.e., inputs and outputs are the sequences, and a term in the output sequence depends on the input sequence and previously produced terms in the output sequence. In NMT, the input sequence is the source language sentence, and the output sequence is the target language sentence. The learning objective is to find the correct target language translation of the source sentence.

Let the equation represent the source sentence:

$$S(s_1; s_2; \ldots; s_n) \tag{4.1}$$

where n is the length of the source sentence.

Then the objective of the NMT is to produce the target sentence:

$$T(t_1; t_2; \ldots; t_m) \tag{4.2}$$

of length m, such that the probability of the target sentence given the source sentence is the maximum.

$$argmax P(T \mid S) \tag{4.3}$$

As a word in a target sentence is dependent not only on the source sentence but also on the previous target words produced, we can write the equation for producing the ith word in the target sentence as:

$$argmax \prod_{i=1}^{m} P(t_i \mid t_{j<i}, S) \tag{4.4}$$

4.4 THE ENCODER-DECODER MODEL

Typically, the models used for NMT are encoder-decoder models (Cho et al., 2014b; Sutskever et al., 2014), consisting of two components: the encoder and the decoder, as illustrated in Figure 4.1.

4.4.1 Encoder

The task of the encoder is to scan through the source sentence one word at a time and convert it into a fixed-length vector, known as a context vector,

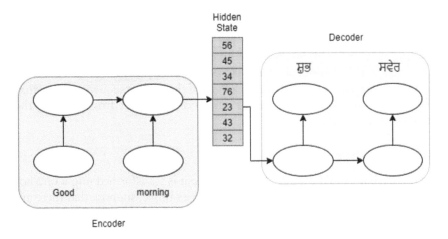

Figure 4.1 The encoder-decoder model.

that encodes the meaning of the sentence. The encoder processes this vector through its hidden layers.

4.4.2 Decoder

The decoder takes the context vector produced by the final hidden layer of the encoder as an input and processes it to produce the target sentence word by word by processing it through its hidden layers. Each neuron has a non-linear activation function (hyperbolic tangent, logistic, Rectified Linear Unit [ReLU]), which can be viewed as a feature detector.

The encoder-decoder model is an end-to-end model that processes the source sentence directly into the target sentence without producing any visible intermediate results. The encoders and decoders are essentially RNNs (Mikolov et al., 2013a, 2013b) or their gated versions (Gated Recurrent Unit, GRU) (Chung et al., 2014, 2015) or LSTM (Hochreiter & Schmidhuber, 1997) capable of learning long-term dependencies.

4.5 RNNs AS ENCODER-DECODER MODELS

As discussed earlier, NMT is a sequence-to-sequence task. In order to model such a task, we need to ensure that:

- the current output is dependent on previous inputs. Also, the model can deal with a variable number of input terms in a sequence,
- the model can share parameters across the sequence, and
- the model can maintain information about the sequence order.

RNNs facilitate dealing with problems involving a temporal aspect. Thus, the initial attempts at modeling machine translation tasks employed RNNs as encoders and decoders (Cho et al., 2014b). It takes an input sequence $X=\{x_1, x_2, ..., x_T\}$ and then processes it recurrently through its hidden layers and produces the output sequence $Y=\{y_1, y_2, ..., y_{T'}\}$, where T is the length of the input sequence, and T' is the length of the output sequence. Note that the number of words in the output sequence can differ from those in the input sequence.

4.5.1 One-hot encoding

Before feeding the data to the RNN, the input is transformed into a one-hot vector, with all but one entry set to zero. For example, suppose that there are ten part-of-speech (POS) tags in English, then the size of the one-hot vector to encode these POS tags is $y \in R^{10}$. The representation of this one-hot vector for nouns can be $y=[1\ 0\ 0\ 0\ 0\ 0\ 0\ 0\ 0\ 0]$; for pronouns, it can be $y=[0\ 1\ 0\ 0\ 0\ 0\ 0\ 0\ 0\ 0]$. Similarly, all the words in a language are encoded as one-hot vectors for machine translation. For this, first, we compute the total number of unique words in a language, say L. Then, we assign a unique identifier between 1 and L to each word. Then, we represent each word using an L-dimensional binary vector with only the bit corresponding to the word identifier set to 1. Figure 4.2 illustrates architecture of the RNN encoder-decoder.

4.5.1.1 Encoder

Encoder (Cho et al., 2014a, 2014b; Sutskever et al., 2014) is an RNN whose state is updated each time it sees a word in a sentence, and the last state of it summarizes the whole sentence, which is called summary vector h_T. The process of encoding a word is as follows.

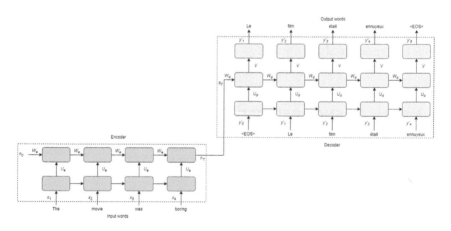

Figure 4.2 The RNN encoder-decoder architecture.

The input word at any time t is fed to the encoder as a one-hot vector x_t. The encoder processes x_t through its hidden layers to create a hidden representation of h_t:

$$h_t = \sigma\left(W_e h_{t-1} + U_e x_t + b_e\right) \tag{4.5}$$

where U_e and W_e are the weight matrices of the encoder, and b_e is the encoder bias, which are the learnable parameters of the model. σ represents a sigmoid activation function. The last hidden state h_T summarizes the entire input sequence.

4.5.1.2 Decoder

The RNN on the decoder (Cho et al., 2014a, 2014b; Sutskever et al., 2014) side takes as input the summary vector h_T generated by the last state of the encoder, the embedding of the previously generated target word, $e\left(\dot{y}_{t-1}\right)$, and the last hidden state of the decoder, s_{t-1}.

$$s_0 = h_T, \tag{4.6}$$

$$s_t = \sigma\left(W_d s_{t-1} + U_d e\left(\dot{y}_{t-1}\right) + b_d\right) \tag{4.7}$$

where U_d and W_d are the weight matrices of the decoder and b_d is the decoder bias.

After processing input, a probability distribution over words in the target language vocabulary is obtained. Target words are then sampled from this probability distribution.

$$P\left(y_t \mid y_1^{t-1}, x\right) = softmax\left(V s_t + c\right) \tag{4.8}$$

4.5.2 Variations of RNNs

RNNs can vary in directionality, learning algorithm used, layer depth, architecture, and inference method, as described below.

4.5.2.1 Directionality

RNN can be uni-directional or bi-directional. Bi-directionality in RNNs was introduced by Schuster and Paliwal (1997). There are two recurrent cells in the bi-directional RNN; one scans the information forward and one in the backward direction. Bi-directional RNN can capture context information, which enhances the translation quality. For example, Bahdanau et al. (2015) and Wu et al. (2016) used bi-directional RNN at the bottom layer. The major disadvantage of using bi-directionality is that it is more time-consuming.

4.5.2.2 Learning

One of the reasons behind successful deep learning models is the development of efficient learning algorithms. These learning algorithms are based on backpropagation (Rumelhart et al., 1986), such as stochastic gradient descent (SGD) (Friedman, 2002), momentum-based SGD, AdaGrad (Duchi et al., 2011), AdaDelta (Zeiler & Fergus, 2014), Adam (Kingma & Ba, 2014), and RMSProp.

4.5.2.3 Depth

Single-layer RNN performs poorly as compared to multi-layer RNN. However, simply increasing the number of layers is not always beneficial. According to Britz et al. (2017), using a four-layer RNN in the encoder for a specific dataset would produce the best performance when there is no other auxiliary method in the whole model. Using many layers may make the network too slow and difficult to train. Increasing the depth also results in vanishing and exploding gradient problems (Bengio et al., 1994), which can be alleviated using LSTMs or GRUs. Wu et al. (2016) suggested residual connections between layers, which improved the backward gradient flow and sped up the convergence process. A deeper network model also indicates a more extended capacity, resulting in poor performance when the training data is less due to over-fitting.

4.5.2.4 NMT architecture selection

There are three common choices for selecting an architecture for NMT: vanilla RNN, LSTM (Hochreiter & Schmidhuber, 1997), and GRU (Cho et al., 2014b). LSTM and GRU are more robust than vanilla RNN when dealing with exploding and vanishing gradients (Bengio et al., 1994).

4.5.3 Discussion and inferences

Greedy search and beam search are the two popular methods for decoding the translation. In a greedy search, the word with the highest probability among the predicted words is output as the translated word at each timestep. The predicted word is then fed to the network for production of the next translated word. Beam search (Britz et al., 2017; Graves, 2012) uses a beam width of size k, which considers the top-k translated words at each timestep. These k-predicted words are fed to the neural network to produce the next translated word. This process is repeated at each timestep until the end of the input sentence is reached.

4.5.3.1 Evaluation metrics

In order to measure the goodness of a machine translation, the most popular evaluation metrics include BLEU (Papineni et al., 2002), METEOR (Lavie & Denkowski, 2009), RIBES (Isozaki et al., 2010), and NIST (Doddington, 2002). The BLEU score compares the given machine-translated text with a reference text and counts the number of matching n-grams between the given machine-translated text and the reference text. BLEU does not consider word order and is language-independent. METEOR scores translations using alignments based on exact, stem, synonym, and paraphrase matches between words and phrases. RIBES is based on rank correlation coefficients modified with precision. NIST is a variation of BLEU, where instead of treating all n-grams equally, weightage is given on how informative a particular n-gram is. Guzmán et al. (2017) proposed using deep learning techniques for the evaluation of machine translations.

4.5.3.2 Advantages of RNNs

NMT systems have several advantages over the existing PBSMT systems (Koehn & Knowles, 2017). The NMT systems do not assume domain knowledge or linguistic features in source and target language sentences. Secondly, the entire encoder-decoder models are jointly trained to maximize the translation quality instead of the PBSMT systems, in which the individual components need to be trained and tuned separately for optimal performance. NMT gathers information from the entire source sentence before translating; as a result, it can capture long-range dependencies in languages, e.g., gender agreements; structural orderings of subject, verb, and object; etc. (Luong, 2017).

4.5.3.3 Limitations of RNNs

The gradients of the RNN are easy to compute via backpropagation through time (Rumelhart et al., 1986; Werbos, 1990), so it may seem that RNNs are easy to train with gradient descent. However, in reality, the relationship between the parameters and the dynamics of the RNN is volatile, making gradient descent ineffective.

While NMT can translate well for short- and medium-length sentences, it has difficulty dealing with long sentences. As the length of the sentence increases, it fails to encode it efficiently in a fixed summary vector, degrading translation performance (Cho et al., 2014b). Using a large size vector will make the RNN slower. Simple RNN uses *tanh* or sigmoid as an activation function, which fails to capture various long-term dependencies in sentences as it puts more emphasis on the latest words seen (Hochreiter, 1998). Also, this suffers from the vanishing and exploding gradients while training (Pascanu et al., 2013). Exploding gradients is the phenomenon in which the gradients become exponentially large as we backpropagate over time, making learning unstable. The problem of exploding gradients can be handled with the help of gradient clipping, as explained in Section 5.9. On the other hand, vanishing gradients are the opposite problem when the gradients go exponentially fast toward zero, turning BPTT into a truncated BPTT that cannot capture long-range dependencies in sequences. LSTMs provide a mechanism to handle the vanishing gradient problem, as explained in Section 6. Moreover, the basic encoder-decoder architecture suffers from the out-of-vocabulary (OOV) problem, which means it cannot correctly translate rare words because of its relatively small vocabulary. OOV words are simply replaced by <unk> symbol in such systems.

4.5.3.4 *Exploding gradient problem: gradient clipping*

The first approach was proposed by Pascanu et al. (2012) in the form of temporal element-wise clipping. At each timestep during backpropagation, any elements greater than a positive threshold τ or smaller than $-\tau$ will be set to τ or $-\tau$, respectively. One can also perform gradient norm clipping, as suggested by Pascanu et al. (2013). Given a final gradient vector g computed per mini-batch, if its norm $\|g\|$ is more significant than a threshold τ, we will use the following scaled gradient $\tau/\|g\|*g$ instead. The latter approach is widely used in many systems nowadays and can also be used in conjunction with the former.

4.6 LSTMS: DEALING WITH LONG-TERM DEPENDENCIES AND VANISHING GRADIENTS

One way to deal with the inability of gradient descent to learn long-range temporal structure in a standard RNN is to modify the model to include "memory" cells using an approach known as "Long Short-Term Memory" (Hochreiter & Schmidhuber, 1997). LSTMs ensure that they selectively process the information so that only important content is retained in a cell. LSTMs have a gating mechanism with three gates to selectively write, read, and forget the information in a sentence. They also have a memory cell to store the contents filtered by these gates. Let s_{t-1} be the previous cell state.

   Selective write: Selective writing is achieved through o_t, the output gate.

$$o_t = \sigma\left(W_o h_{t-1} + U_o x_t + b_o\right) \tag{4.9}$$

where W_o and U_o are the weight matrices for the output gate and b_o is the output gate bias. The hidden state h_t is then computed using a selective write as follows:

$$h_t = o_t \odot s_t \tag{4.10}$$

Selective read: Let $\dot{s}_t$ be the current (temporary) cell state, which can be written as follows:

$$\dot{s}_t = \sigma\left(Wh_{t-1} + Ux_t + b\right) \tag{4.11}$$

$\dot{s}_t$ captures all the information from the previous state h_{t-1} and current input x_t. However, we may not want to use all this new information and only selectively read from it before constructing the new cell state. This is achieved using i_t the input gate.

$$i_t = \sigma\left(W_i h_{t-1} + U_i x_t + b_i\right) \tag{4.12}$$

where W_i and U_i are the weight matrices for the input gate and b_i is the input gate bias.

Selective forget: f_t is the forget gate which is computed as:

$$f_t = \sigma\left(W_f h_{t-1} + U_f x_t + b_f\right) \tag{4.13}$$

where W_f and U_f are the weight matrices for the forget gate, b_f is the forget gate bias.

We can now store the content in a cell using the following equation:

$$s_t = f_t \odot s_{t-1} + i_t \odot \dot{s}_t \tag{4.14}$$

4.6.1 GRUs

LSTMs have many variants, including different numbers and different arrangements of gates. GRU is one of the popular variants of LSTMs. GRUs have two gates: the output gate o_t and the input gate, which is computed as:

$$o_t = \sigma\left(W_o s_{t-1} + U_o x_t + b_o\right) \tag{4.15}$$

$$i_t = \sigma\left(W_i s_{t-1} + U_i x_t + b_i\right) \tag{4.16}$$

The cell states $\grave{s}_t$ and s_t are computed as:

$$\grave{s}_t = \sigma\left(W\left(o_t \odot s_{t-1}\right) + Ux_t + b\right) \tag{4.17}$$

$$s_t = \left(1 - i_t\right) \odot s_{t-1} + i_t \odot \grave{s}_t \tag{4.18}$$

4.6.2 Limitations of LSTMs

When the length of the input sentence becomes significant, even the LSTMs start to show signs of strain (Van Houdt et al., 2020). The model tends to forget the contents of the distant word positions when the sentences are long. This means that retaining the context of a distant word decreases exponentially with increasing distance. Another problem with RNNs and LSTMs is that it is hard to parallelize the work of processing sentences since they have to be processed word by word. Moreover, there is no model for capturing long- and short-range dependencies.

To summarize, LSTMs and RNNs present the following problems:

- No explicit modeling of long- and short-range dependencies.
- istance" between positions is linear.

4.7 NMT WITH ATTENTION

The encoder-decoder architecture was almost immediately extended by Bahdanau et al. (2014), who proposed a soft-search model that uses an attention mechanism in the encoder-decoder model to address the sentence length problem (refer to Figure 4.3). While producing a translation, it focuses selectively on the most relevant portions of the source sentence. Instead of using a fixed-length vector to represent a sentence, this mechanism represents each word with a fixed-length vector called an annotation vector. The decoder is provided only with relevant information at each step and is not overloaded with irrelevant information, enhancing system performance. The only difference with the basic encoder-decoder is in the decoder part; the encoder part is the same in both models. In the basic model, the decoder takes an encoded summary vector as input, but in the attention model, it takes a context vector as input to generate the target sentence. Luong et al. (2015a) later extended this work by proposing some other improvements.

As in the basic encoder-decoder model, the last hidden state of the encoder, i.e., the summary vector, h_T becomes the input s_0 to the decoder.

$$s_0 = h_T \tag{4.19}$$

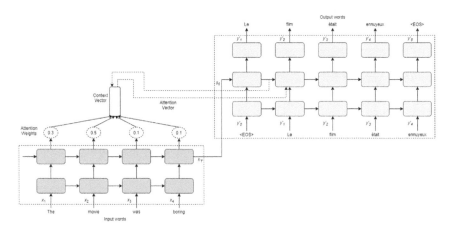

Figure 4.3 The attention mechanism.

Let the decoder RNN's state at timestep t be s_t, and h_j be the encoder RNN's state at step j. Then, we define a function e_{jt}, the alignment score, as:

$$e_{jt} = f_{ATT}\left(s_{t-1}, h_j, \theta\right) \tag{4.20}$$

This means that e_{jt} at timestep t depends on the previous state of the decoder, s_{t-1}, the encoder hidden state, h_j for the jth input word ($\forall j, j \in \{1, 2, \cdots, T\}$), and the model parameters, θ. This quantity captures the importance of the jth input word for decoding the tth output word. In other words, e_{jt} captures the attention weight scores by comparing the decoder state s_{t-1} with all the encoder hidden states h_j. For example, e_{j1} is calculated for the decoder state at timestep 1 as:

$$e_{11} = f\left(s_0, h_1\right), e_{21} = f\left(s_0, h_2\right), \ldots, e_{T1} = f\left(s_0, h_T\right) \tag{4.21}$$

Thus,

$$e_1 = \left[e_{11}, e_{21}, \ldots, e_{T1}\right] \tag{4.22}$$

From the above equation, we can say that:

$$e_t = \left[e_{1t}, e_{2t}, e_{3t}, \ldots, e_{Tt}\right] \tag{4.23}$$

e_{jt} could be defined in different ways; for example,

$$e_{jt} = \begin{cases} h_j W s_{t-1} \; [Luong's\,version] \\ V_{att}^T tanh\left(W_{att} h_j + U_{att} s_{t-1}\right) [Bahdanau's\,version] \end{cases} \tag{4.24}$$

We can normalize these weights to get attention weights, α_{jt}, by using the softmax function as:

$$\alpha_{jt} = \frac{exp(e_{jt})}{\sum_{j=1}^{M} exp(e_{jt})} \tag{4.25}$$

where α_{jt} denotes the probability of focusing on the jth input word to produce the tth output word. Thus, the attention weights for generating the output word at time step t can be written as:

$$\alpha_t = [\alpha_{1t}, \alpha_{2t}, ..., \alpha_{Tt}] \tag{4.26}$$

Using the attention weights, α_{jt}, we can now define the context vector, c_t, which is a weighted average of the encoder hidden states h_j as:

$$c_t = \sum_{j=1}^{T} \alpha_{jt} h_j \tag{4.27}$$

The final attention vector is generated by concatenating the context vector with the current decoder hidden state. The current decoder hidden state, s_t, is now a non-linear function of the previous decoder hidden state, s_{t-1}, and the final attention vector as:

$$s_t = RNN\left(s_{t-1}, \left[e\left(\dot{y}_{t-1}\right), c_t\right]\right) \tag{4.28}$$

Then a probability distribution over the target vocabulary and the sample target word is calculated, as done in the basic encoder-decoder model, until the completion of the translation.

The attention mechanism discussed above is known as global attention, which calculates the context vector by attending to all the words in the source sentence. The major drawback of global attention is that the network becomes very slow when a large input sentence is involved (Luong et al., 2015a). The researchers have also tested other variations of attention mechanisms, such as local attention and other attention strategies like the dot product, linear combination, etc.

Local attention: First proposed by Luong et al. (2015a), local attention alleviates the extensive computation of the context vectors by limiting the scope of the attention vector. Given the position of the current target word p_t, a scope D is defined such that the context vector c_t is derived as a weighted average over the set of source hidden states within the range $[p_t - D; p_t + D]$.

4.8 RECENT DEVELOPMENTS IN NMT

Allen (1987) and Chrisman (1991) wrote the very first papers on encoder-decoder models for translation. But due to the limited computing power and limited data at that time, research in the field almost stagnated for many years. Schwenk (2012) first demonstrated the use of feed-forward neural networks to directly learn the translation probability of phrase pairs for SMT using continuous representations on the English/French IWSLT task. Devlin et al. (2014) also used a similar approach and proposed a feed-forward neural network to model a translation model. Their network could predict only a single target word at a time and was also constrained by an a priori fixed maximum input phrase length.

The landmark paper demonstrating the successful application of deep learning to the task of machine translation was proposed by Kalchbrenner and Blunsom (2013b). They used a Recurrent Language Model based on RNNs to generate the translation, conditioned on the source sentence, using a convolutional sentence model to translate English sentences to French. The authors were the first to map sentences to vectors and then back to sentences. However, they lost the ordering of words due to CNNs for mapping (Sutskever et al., 2014). Cho et al. (2014b) proposed an encoder-decoder framework for sequence-to-sequence modeling. They proposed using the final state of the encoder's hidden layer as the source sentence's encoding. Sundermeyer et al. (2014) used a bi-directional RNN in the encoder and used the concatenation of the hidden layers' final states as the source sentence's encoding.

Sutskever et al. (2014) proposed the first machine translation model entirely based on neural networks. The network used an RNN (LSTM or GRU). They trained the encoder using the source sentence in the reverse order of words and the decoder in the correct word order of the target sentence. The model achieved accuracy comparable to that of state-of-the-art SMT systems. Their approach could not use a fixed-size vector for the entire sentence, with limited representation capability when sentences become longer.

Usually, encoder-decoder models deploy multiple layers stacked on top of each other (Luong et al., 2015a; Sutskever et al., 2014). Bahdanau et al. (2014), Cho et al. (2014b), and Jean et al. (2015a) all adopted a different version of the RNN with an LSTM-inspired remote unit, GRU, for both components. They all used a single RNN layer except for the latter two works, which utilized a bi-directional RNN for the encoder. Wu et al. (2016) proposed GNMT, which marked the shift of Google from SMT to RNN-based NMT.

Besides the work mentioned above, other researchers have also proposed different architectures with excellent performance. Zhou et al. (2016) designed linear connections called fast-forward connections using LSTM, which enabled building a deeper network for better performance. Shazeer et al. (2017) proposed a sparsely gated mixture of experts (MoE) layer into the GNMT model, consisting of multiple feed-forward sub-networks. A gate function governs a sparse combination of these sub-networks connected with the RNN layer. Although it generally requires more parameters, this method has outperformed the original GNMT model.

4.8.1 Word embeddings

Before feeding the data to the neural network, there is a pre-processing task of converting the words to vectors called word embeddings. Word embedding maps words in the input sentences into continuous space vectors. Each value in the vector weighs the input word from a different perspective. Words with similar meanings tend to have similar word embeddings. For example, the word embeddings for "king" and "queen" may have the same value for the dimension "royalty", while "king" and "man" might be related along the dimension "gender." Thus, word embeddings can capture semantic and contextual information.

Before the emergence of transformer-based models, word2vec (Mikolov et al., 2013a, 2013b) and GloVe (Global Vectors for Word Representation) proposed by Pennington et al. (2014) were the primary approaches for creating word embeddings, which were pre-trained on large amounts of unlabeled text. Word2vec is based on the Continuous Bag of Words (CBOW) and skip-gram models. CBOW guesses the target word given the surrounding context words across a window of size k in both the left and right directions. On the other hand, the skip-gram model predicts the surrounding context words given the central target word. The GloVe is a global log-bilinear regression model aggregating the global word-word co-occurrence matrix from a corpus to produce word embeddings. The GloVe is a count-based method that counts the co-occurrence of the context word and the target word within a window of size n.

The representations produced by word2vec and GloVe are considered context-independent representations. The embedding learned for a word does not change depending on the context. For example, the embedding learned for the word "bark" remains the same, regardless of whether the word "bark" is used in the phrase "the dog's bark" or in the phrase "the tree bark." Another limitation of these approaches was that the word embeddings generated were shallow representations that could only be used in the first layer of the neural network. The rest of the network had to be trained from scratch.

Another approach, Embeddings for Language Models (ELMo), proposed by Peters et al. (2018), learns bi-directional word embeddings using LSTMs. ELMo produces context-dependent representations, unlike word2vec and GloVe. It uses two LSTMs: one processes the information in the forward direction, and the other processes it in the backward direction to model the context on both the left and right sides of a word. The bi-directional LSTMs are pre-trained on a large text corpus with a language modeling objective. The output of the last hidden layer is then concatenated to produce the word embeddings.

4.8.2 CNN-based NMT

Several researchers also tried to use CNNs for the task of machine translation, but attention-based encoder-decoder models soon outperformed these models. For example, Kalchbrenner and Blunsom (2013a) used a CNN encoder with an RNN decoder. Cho et al. (2014b) used a gated recursive CNN encoder and an RNN decoder, but the CNN encoder's performance was worse than the RNN encoder. Kaiser and Bengio (2016) proposed an extended model of active memory akin to existing attention-based NMT models using an extended neural GPU (Kaiser & Sutskever, 2015). Gehring et al. (2017) achieved the best performance among CNN-based NMT models using a CNN encoder and got a performance comparable to that of the RNN-based model. Kalchbrenner et al. (2016) proposed a CNN-based NMT model called ByteNet. It achieved good performance on character-level translation but failed at word-level translation.

CNN-based models are faster as compared to RNN-based models. This is due to the inherent structure of the CNNs, which allows for parallel computations. Moreover, solving the vanishing gradient problem in CNN-based models is easier. The limitation of CNN-based models is that they can only capture word dependencies within the filter width. Thus, the long-range dependencies can only be captured at higher convolution layers, making CNNs perform worse than RNNs in the machine translation task. Earlier CNN-based models could also not handle longer sentences using fixed-length vectors.

Gehring et al. (2017) proposed an attention-based convolutional NMT model that overcame the fixed-context limitation of the original CNN-based NMT models. They stacked multiple convolutional layers for the encoder and decoder. Each layer consisted of one-dimensional convolutions followed by a gated linear unit (GLU) (Dauphin et al., 2017). The output of the current decoder layer and the output of the final encoder layer are used to compute the dot-product attention. Positional information is retained with the help of positional embeddings. Although this approach yielded a better result than the RNN-based NMT models, it soon outperformed the transformer-based models (Vaswani et al., 2017).

4.8.3 Fully attention-based NMT

In the earlier approaches to NMT, the attention mechanism was used as a supplementary component to deal with the long-range dependencies within the sentences (Bahdanau et al., 2014). Later, Vaswani et al. (2017) proposed a fully attention-based NMT model, known as the transformer. The transformer model is based on self-attention, a mechanism that generates a representation of the input sequence by relating the different positions (Cheng et al., 2016a). The transformer allows reading the entire sentence parallelly instead of reading it sequentially word by word, as in the case of RNN-based NMT models. This helps in increasing the inference speed of the network. The transformer model can address the problems encountered in the previous NMT models. It is designed to achieve parallel computations, unlike RNN-based NMT models, in which the temporal dependencies limit the speed of computations. It also addresses the limited context problem encountered in CNN-based NMT models through self-attention.

Self-attention generates an attention-based vector representation of the input sequence by calculating the word dependencies inside the sequence. Self-attention calculates the query vector, Q, the key vector, K, and the value vector, V, for each input word embedding. These vectors are learned during the training process. The attention vector for the target word is calculated by taking the dot product of the query vector Q of the target word embedding and the key vector K of the previous input word embeddings. The vector thus obtained is then multiplied by the value vector V of the input word embeddings. Formally, the attention weights are calculated as:

$$Attention(Q, K, V) = softmax\left(\frac{QK^T}{\sqrt{d_k}}\right)V \qquad (4.29)$$

where $\frac{1}{\sqrt{d_k}}$ is a scaled factor (the dimension of the embeddings) to avoid more stable gradients caused by dot-product operations.

The transformer network consists of an encoder transformer and a decoder transformer. The encoder (decoder) transformer consists of multiple identical encoder (decoder) units stacked on top of each other. The order of the input words is captured using positional encoding vectors. Each encoder unit consists of a self-attention layer with multi-head attention, followed by a feed-forward network. Multi-head attention is used to project each input word embedding onto different subspaces for each query by producing multiple attention heads. The output of these attention heads is then concatenated and transformed linearly. Each decoder unit has three layers: masked multi-head attention, cross-attention, and feed-forward. Cross-attention allows the decoder to attend to parts of the input sequence. The encoder transformer and the decoder transformer contain the same number of units. The total number of units used in a transformer network is a hyperparameter. Vaswani et al. (2017) employed six units (refer to Figure 4.4).

The original transformer model had the limitation that it could handle only fixed-length input sequences. Therefore, longer sentences had to be split into several small segments. This resulted in the problem of context fragmentation, which means that the context was lost among the segments.

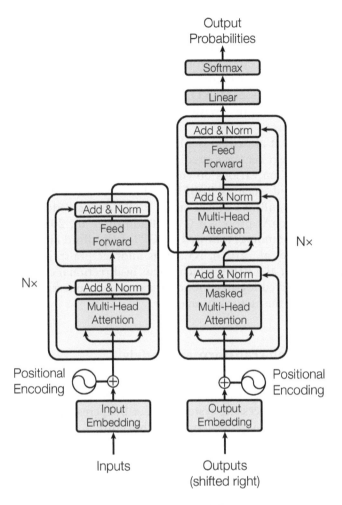

Figure 4.4 The attention network (Vaswani et al., 2017).

4.8.4 Transformer-based pre-trained models

Pre-trained models are those models that have already been trained on an enormous amount of unlabeled text. Using a labeled dataset, these models can be fine-tuned for downstream tasks such as machine translation, sentiment analysis, text summarization, question-answering, etc. Pre-trained models eliminate the need to build a model from scratch. Pre-trained models can be autoregressive or autoencoding. Autoregressive models are trained on the language modeling task, where the input is processed from left to right and the next token is predicted considering all the previous tokens. Generative pre-trained transformers (GPTs), proposed by Radford et al. (2018a), is such a model. In autoencoding models, the input tokens are corrupted somehow, and the model then attempts to reconstruct the original input sequence. These are also known as masked language models (MLMs), such as BERT proposed by Devlin et al. (2019), ALBERT proposed by Lan et al. (2020), etc.

Tech giants have developed many pre-trained models like Google AI, Facebook AI Research, OpenAI, etc. These models are trained on heavy amounts of text with many parameters, transformer blocks, and hidden layers.

BERT (Devlin et al., 2019) is a bi-directional language model that uses an MLM and next sentence prediction (NSP) for pre-training. After the success of BERT on various downstream tasks, many improved variants of BERT emerged, such as RoBERTa (Liu et al., 2019), ALBERT (Lan et al., 2020), SpanBERT (Joshi et al., 2020), StructBERT (Wang et al., 2019b), etc. OpenAI proposed GPT (Radford et al., 2018a) for natural language understanding, followed by GPT-2 (Radford et al., 2018b) and GPT-3 (Brown et al., 2020). XLNet is another pre-trained model proposed by Yang et al. (2019), which learns bi-directional context by maximizing the expected log-likelihood of a sequence concerning all possible permutations of the factorization order.

4.8.5 Improved transformer models

Since the origin of the vanilla transformer model, a variety of pre-trained transformer-based architectures have been proposed to push the limits of natural language processing systems, either by increasing the depth of the network or by modifying the network architecture. Bapna et al. (2018) proposed a 2–3 times deeper encoder Transformer for machine translation by modifying the attention mechanism to extend its connection along with the encoder depth, similar to weighted residual connections. Wang et al. (2019a) continue along the same line by proposing a 30/25 layer encoder in the Transformer model combined with layer normalization and a dynamic linear combination of layers to pass the features extracted from the preceding layers to the next.

Dehghani et al. (2019) proposed a universal transformer bagged with a technique called the adaptive computation time halting mechanism, which allows the model to apply the encoder and decoder a variable number of times for each symbol. So et al. (2019) used tournament selection-based neural architecture search to discover a more efficient transformer architecture by developing the progressive dynamic hurdles method and came up with the Evolved Transformer.

Shaw and Uszkoreit (2018) tried to improve the representations of the positional information of the input tokens by extending the self-attention to consider the pairwise relationships between the input tokens by modeling the input as a labeled, directed, fully connected graph.

4.9 NMT IN LOW-RESOURCE LANGUAGES

Languages such as English, Spanish, German, etc. have sufficient resources, including datasets and techniques, for performing research in the field of machine translation. NMT models based on deep learning are resource-hungry and require billions of parallel sentences for training. This is why most NMT systems involving high-resource languages exhibit high performance at par with human translation. Low-resource languages like Hindi or Punjabi, on the other hand, do not possess sufficiently large parallel corpora, and thus, the translation performance of NMT systems on these languages is relatively poor compared to PBSMT systems (Koehn & Knowles, 2017).

In the absence of large parallel corpora in low-resource settings, the researchers primarily leverage monolingual corpora, which are sufficiently available, via unsupervised training. Lample et al. (2020) argue that unsupervised NMT outperforms supervised NMT if a few additional data resources are available. Back-translation, initially proposed by Edunov et al. (2020) and Sennrich et al. (2016), has been heavily explored in low-resource environments where bilingual parallel data is augmented by generating synthetic parallel data. In particular, an NMT system outputs the translations of the abundantly available target-side monolingual data. These translations are then added to the source-side data. Back-translation has been the standard approach for unsupervised NMT (Artetxe et al., 2017; Lample et al., 2017, 2020). Sennrich et al. (2016) relied on monolingual data for Turkish-English language pairs and used back-translation from target to source to generate additional bilingual training data.

Lample et al. (2020) propose to rely only on large monolingual corpora in each language using unsupervised machine translation through careful initialization of model parameters, training language models in both source and target languages, iterative back-translation for automatic generation of parallel sentences, and sharing of encoder parameters across the source and target languages. They proposed two model variants: a neural-based model and a phrase-based model. The authors achieved 28.1 and 25.2 BLEU points on the WMT'14 English-French and WMT'16 German-English benchmark datasets without any parallel data. They achieved better results than supervised and semi-supervised approaches on low-resource language pairs such as English-Urdu and English-Romanian. They also showed that phrase-based models outperformed neural-based models in a fully unsupervised environment.

Cheng et al. (2016b) also exploited monolingual corpora and proposed a semi-supervised approach based on parallel and monolingual data concatenation. Specifically, they used an autoencoder to reconstruct the monolingual corpora by encoding the target sentence into the source sentence and then decoding the source sentence to reconstruct the target sentence. They experimented with Chinese-English language pairs.

Ahmadnia and Dorr (2019) argue that although data augmentation techniques such as back-translation enlarge the bilingual parallel corpora, there is no quality control as the synthetic data may contain errors and noise. When this data is used further during the iterative training process, such errors will be reinforced, and thus the accuracy of the synthetic parallel dataset cannot be guaranteed. Similarly, Sennrich and Zhang (2019) emphasize that although semi-supervised and unsupervised approaches yielded good results on some language pairs, their performance depended on the availability of large amounts of auxiliary data and other conditions being met. For example, unsupervised methods do not perform well when the language pairs are morphologically different or when the training domains do not match (Søgaard et al., 2018).

Therefore, some researchers tackled the problem of low-resource NMT without using any auxiliary data. For example, Östling and Tiedemann (2017) introduced more local dependencies and used word alignments to learn sentence reordering during translation. Nguyen and Chiang (2018) used smaller and fewer layers for smaller datasets. Sennrich and Zhang (2019) emphasized that one can achieve satisfactory results through careful initialization of parameters and other techniques such as bi-directional deep RNNs, label smoothing, word dropout, layer normalization, and tied embedding rather than using a large amount of auxiliary data for low-resource language pairs. They also pointed out that small batch sizes benefit low-resource NMT. They experimented on German-English and Korean-English language pairs and got 10.37 BLEU points on the latter language pair, thus improving by 4 BLEU points on the previously reported results. The authors also showed that NMT beats PBSMT with as little as 10,000 words of parallel training data using their approach.

Ren et al. (2019) tried to eliminate noise in the synthetic data by using SMT models as posterior regularizers. Then they jointly optimize the SMT and NMT models to boost each other incrementally in a unified expectation-maximization framework. Some researchers have also experimented with parallel data involving other language pairs (Cheny et al., 2017; Neubig & Hu, 2018; Nguyen & Chiang, 2017; Zoph et al., 2016).

Among other approaches, Gulcehre et al. (2015) integrated a separately trained language model into the target language for a low-resource English-Turkish scenario. Specifically, they integrated the language model into the training parts of an NMT model with additional objectives, including a language modeling objective. Gu et al. (2018) proposed a model-agnostic meta-learning (MAML) algorithm wherein they employ multiple high-resource language pairs to find the initial parameters for a language model for low-resource language pairs, which can then be trained only with a few examples. For example, the authors initialized the model using 18 high-resource language pairs, and they were able to achieve 22.04 BLUE points on new language pairs trained only on 600 parallel examples.

The translation quality of NMT models for low-resource languages is inferior to that of high-resource languages. Specifically, the NMT models for such languages perform poorly on domain-specific data, conflating different dialects of a language, producing overly literal translations, and having poor informal and spoken language performance.

Active learning (AL), first proposed by Settles (2012), can be a promising method to deal with the challenges of low-resource machine translation. AL is an interactive learning technique where the user can be queried to label the data selectively. The data to be labeled is selected automatically and iteratively by the AL methods so that the most useful examples get labeled to maximize the model's performance and minimize the labeling cost. The model is then trained on labeled data and tested on unlabeled data. The labels output by the model are then passed on to the AL methods, which again present new unlabeled instances to the user based on this previous output, and this process is iterated until the model achieves the desired performance.

4.10 VOCABULARY COVERAGE PROBLEM

In NMT systems, there is a trade-off between the vocabulary size and the system's performance. NMT systems try to cover the frequently used words in a language, leaving out the rarely used words. This saves computational costs but may deteriorate the translation quality. The rare words unknown to the NMT system are called out-of-vocabulary (OOV). Whenever an NMT system encounters OOV words, they are replaced by the "UNK" symbol. The "UNK" symbol in a sentence is semantically incomplete and may lead to ambiguity when it replaces a critical word (Gulcehre et al., 2016). Thus, OOV words affect the translation quality of an NMT system.

Several kinds of solutions have been proposed in the literature to handle OOV words. One is based on finding methods to increase the computational speed of NMT systems so that they can handle larger vocabulary sizes; for example, Jean et al. (2015b) propose to use a considerable target vocabulary size without increasing the computational complexity by using a method based on importance sampling. Another approach relies on transliterating named entities and some of the rare words to the target language. Some researchers have directly translated the OOV words into the target language through a dictionary lookup, for example, the "copy" mechanism (Luong et al., 2015b). Gu et al. (2016) proposed a CopyNet model that used a copying mechanism to select sub-sequences in the source sentence and then place them in the target sentence. The authors claimed that although CopyNet is a general mechanism to carry information to the next stage, it has proved beneficial in solving the OOV problem.

Among the other advanced approaches, Gulcehre et al. (2016) proposed using a pointer softmax to handle OOV words. Specifically, they used two software layers to predict the next target word: one for predicting the location of the corresponding word in the source sequence, called the location softmax, and the other for generating the target word from a shortlist vocabulary, called the shortlist softmax. They use a switching network built using a multi-layer perceptron to decide which of the two softmax layers to use.

Byte-pair encoding proposed by Sennrich et al. (2015) used a data compression method called Byte-Pair Encoding (BPE), through which they extracted sub-words. Chung et al. (2016) also used BPE to extract sub-word symbols from the source sentence. They used both BPE-based sub-word symbols and a sequence of characters on the target side.

Character-level machine translation has also received attention as a mechanism for handling OOV words. Costa-Jussa and Fonollosa (2016) deployed a CNN with a highway network on the encoder side for character modeling, while on the decoder side, they generated the target words. Luong and Manning (2016) proposed using a combination of word-level and character-level NMT models using RNNs, which uses the word-level translations under normal circumstances and refers to the character-level RNN only when it encounters an OOV word. Lee et al. (2017) and Ling et al. (2016) also used a character-level NMT model based on RNNs.

More recently, Luong and Manning (2016) proposed a hybrid model that combines the word-level RNN with the character-level RNN for assistance, translating mostly at the word level and consulting the character-level RNN when it encounters an OOV word.

Sennrich and Haddow (2016) proposed incorporating linguistic features and the input sequence into the NMT system to improve the NMT output. Using linguistic features in NMT is also known as factored NMT. Linguistic features such as lemma, POS tags, dependency parsing labels, sub-word tags, etc. can be added at the source and target sides. The authors have shown improvements in the NMT system by using linguistic features. The main drawback of factored NMT is that the datasets annotated with linguistic features are not available for most language pairs.

4.11 DATASETS FOR MACHINE TRANSLATION

A large number of datasets have been/are being developed for the task of machine translation. Among the most worth mentioning is the Tatoeba open project, Tatoeba (2021), which supports 409 languages. The tab-delimited bilingual sentence pairs between English and other languages can be accessed at *Tab-Delimited Bilingual Sentence Pairs*, 2021. CLARIN ERIC (European Research Infrastructure Consortium) infrastructure

Table 4.1 Datasets for NMT research

Dataset source	URL
Tatoeba project	https://tatoeba.org/en/
CLARIN ERIC	https://www.clarin.eu/resource-families/parallel-corpora
CCMatrix	https://github.com/facebookresearch/LASER/tree/master/tasks/CCMatrix
WikiMatrix	https://github.com/facebookresearch/LASER/tree/master/tasks/WikiMatrix
WMT'19 datasets	http://www.statmt.org/wmt19/translation-task.html#download

(*CLARIN Parallel Corpora*, 2021) makes 86 parallel corpora, among which 47 are bilingual corpora consisting of primarily European and some non-European languages such as Hindi, Tamil, and Vietnamese. There are 39 multilingual corpora, with five of them consisting of parallel sentences in more than 50 languages. CCMatrix by Schwenk et al. (2019b) is a collection of 4.5 billion parallel sentences in 576 language pairs pulled from the snapshots of the CommonCrawl public dataset. WikiMatrix (Schwenk et al., 2019a) is another parallel collection of 135 million parallel sentences for 1,620 different language pairs in 85 different languages created using publicly available Wikipedia articles.

The various sources of datasets available for research are summarized in Table 4.1.

4.12 CHALLENGES AND FUTURE SCOPE

With deep learning-based NMT systems, various language pairs have achieved near-human translation results, especially in high-resource languages. Nevertheless, dealing with low-resource languages has been daunting for the NMT research community. There are approximately 700 different languages in the world. Digitalizing these languages will play an essential role in preserving them for future generations. The ability to translate easily from one language to another is essential. Thus, a lot can be done for machine translation of low-resource languages, including developing resources such as parallel corpora, bilingual dictionaries, annotated datasets, etc.

Other challenges include handling rare words, dealing with large vocabularies, developing optimization algorithms, better regularization strategies, optimal learning rates, etc. Another research direction is to explore better deep learning architectures for sequence modeling, for example, temporal convolutional networks (TCNs) (Bai et al., 2018). Multimodal translation is another exciting area of research. Newer techniques such as memory-augmented neural networks (MANNs) and active learning show impressive results. The potential of MANNs has not yet been fully explored, and the research in this direction is still in its infancy, but it promises an exciting future in the field of NMT.

REFERENCES

Ahmadnia, B., & Dorr, B. J. (2019). *Bilingual low-resource neural machine translation with round-tripping: The case of Persian-Spanish. Proceedings of the International Conference Recent Advances in Natural Language Processing, RANLP (2019-September)*, 18–24. https://doi.org/10.26615/978-954-452-056-4_003

Allen, R. B. (1987). *Several studies on natural language and back propagation. IEEE First International Conference on Neural Networks, 2*, 335–341.

Arnold, D. J., Balkan, L., Meijer, S., Humphreys, R. L., & Sadler, L. (1993). *Machine Translation: An Introductory Guide*. Blackwells-NCC, London. https://www.essex.ac.uk/linguistics/clmt/MTbook/

Artetxe, M., Labaka, G., & Agirre, E. (2017). *Learning bilingual word embeddings with (almost) no bilingual data. ACL 2017-55th Annual Meeting of the Association for Computational Linguistics, Proceedings of the Conference (Long Papers), 1*, 451–462. https://doi.org/10.18653/v1/P17-1042

Bahdanau, D., Cho, K., & Bengio, Y. (2014). *Neural machine translation by jointly learning to align and translate. ArXiv E-Prints*, arXiv:1409.0473. https://arxiv.org/abs/1409.0473

Bahdanau, D., Cho, K., & Bengio, Y. (2015). *Neural machine translation by jointly learning to align and translate. 3rd International Conference on Learning Representations, (ICLR)*. https://arxiv.org/abs/1409.0473

Bai, S., Kolter, J. Z., & Koltun, V. (2018). *An empirical evaluation of generic convolutional and recurrent networks for sequence modeling. ArXiv E-Prints, arXiv:1803.*

Bapna, A., Chen, M. X., Firat, O., Cao, Y., & Wu, Y. (2018). *Training deeper neural machine translation models with transparent attention. Proceedings of the 2018 Conference on Empirical Methods in Natural Language Processing*, 3028–3033. https://doi.org/10.18653/v1/D18-1338

Bengio, Y., Simard, P., & Frasconi, P. (1994). Learning long-term dependencies with gradient descent is difficult. *IEEE Transactions on Neural Networks, 5*(2), 157–166. https://doi.org/10.1109/72.279181

Bojar, O., Chatterjee, R., Federmann, C., Graham, Y., Haddow, B., Huck, M., Jimeno Yepes, A., Koehn, P., Logacheva, V., Monz, C., Negri, M., Névéol, A., Neves, M., Popel, M., Post, M., Rubino, R., Scarton, C., Specia, L., Turchi, M., ... Zampieri, M. (2016). *Findings of the 2016 Conference on Machine Translation. Proceedings of the First Conference on Machine Translation: Volume 2, Shared Task Papers*, 131–198. https://doi.org/10.18653/v1/W16-2301

Britz, D., Goldie, A., Luong, M. T., & Le, Q. V. (2017). *Massive exploration of neural machine translation architectures. EMNLP 2017- Conference on Empirical Methods in Natural Language Processing, Proceedings*, 1442–1451. https://doi.org/10.18653/v1/d17-1151

Brown, P. F., Cocke, J., Pietra, S. A. Della, Pietra, V. J. Della, Jelinek, F., Lafferty, J. D., Mercer, R. L., Roossin, P. S., & Watson, T. J. (1990). A statistical approach to machine translation. *Computational Linguistics, 16*(2), 79–85.

Brown, T. B., Mann, B., Ryder, N., Subbiah, M., Kaplan, J., Dhariwal, P., Neelakantan, A., Shyam, P., Sastry, G., Askell, A., Agarwal, S., Herbert-Voss, A., Henighan, T., Chen, M., Ziegler, D. M., Krueger, G., Hesse, C., & Mccandlish, S. (2020). Language models are few-shot learners. ArXiv E-Prints, arXiv:2005.14165.

Cheng, J., Dong, L., & Lapata, M. (2016a). *Long short-term memory networks for machine-reading. EMNLP 2016- Conference on Empirical Methods in Natural Language Processing, Proceedings*, 551–561. https://doi.org/10.18653/v1/d16-1053

Cheng, Y., Xu, W., He, Z., He, W., Wu, H., Sun, M., & Liu, Y. (2016b). *Semi-supervised learning for neural machine translation. 54th Annual Meeting of the Association for Computational Linguistics, ACL 2016-Long Papers*, 4, 1965–1974. https://doi.org/10.18653/v1/p16-1185

Cheny, Y., Liuz, Y., Cheng, Y., & Li, V. O. K. (2017). *A Teacher-Student Framework for zero-resource neural machine translation. Proceedings of the 55th Annual Meeting of the Association for Computational Linguistics (Volume 1: Long Papers)*, 1925–1935. https://doi.org/10.18653/v1/P17-1176

Cho, K. C., Merriënboer, B. van, Bahdanau, D., & Bengio, Y. (2014a). *On the properties of neural machine translation: encoder-decoder approaches. Proceedings of SSST-8, Eighth Workshop on Syntax, Semantics and Structure in Statistical Translation*, 103–111.

Cho, K., Merrienboer, B. van, Gulcehre, C., Bahdanau, D., Bougares, F., Schwenk, H., & Bengio, Y. (2014b). *Learning phrase representations using rnn encoder-decoder for statistical machine translation. Proceedings of the 2014 Conference on Empirical Methods in Natural Language Processing (EMNLP)*, 1724–1734.

Chorowski, J., Bahdanau, D., Cho, K., & Bengio, Y. (2014). *End-to-end continuous speech recognition using attention-based recurrent NN: first results. NIPS 2014 Workshop on Deep Learning*, December 2014.

Chrisman, L. (1991). Learning recursive distributed representations for holistic computation. *Connection Science*, 3(4), 345–366. https://doi.org/10.1080/095 40099108946592

Chung, J., Cho, K., & Bengio, Y. (2016). A character-level decoder without explicit segmentation for neural machine translation. *54th Annual Meeting of the Association for Computational Linguistics, ACL 2016- Long Papers*, 3, 1693–1703. https://doi.org/10.18653/v1/p16-1160

Chung, J., Gulcehre, C., Cho, K., & Bengio, Y. (2014). Empirical evaluation of gated recurrent neural networks on sequence modeling. *NIPS 2014 Workshop on Deep Learning*, December 2014.

Chung, J., Gulcehre, C., Cho, K., & Bengio, Y. (2015). *Gated feedback recurrent neural ntworks*. In F. Bach & D. Blei (Eds.), *Proceedings of the 32nd International Conference on Machine Learning* (Vol. 37, pp. 2067–2075). PMLR. https://proceedings.mlr.press/v37/chung15.html

CLARINParallel Corpora. (2021). https://www.clarin.eu/resource-families/parallel-corpora

Collobert, R., Weston, J., Bottou, L., Karlen, M., Kavukcuoglu, K., & Kuksa, P. (2011). Natural language processing (almost) from scratch. *Journal of Machine Learning Research*, 12(76), 2493–2537. https://jmlr.org/papers/v12/collobert11a.html

Costa -Jussa, M. R., & Fonollosa, J. A. R. (2016). Character-based neural machine translation. *Proceedings of the 54th Annual Meeting of the Association for Computational Linguistics*, 357–361.

Dauphin, Y. N., Fan, A., Auli, M., & Grangier, D. (2017). Language modeling with gated convolutional networks. *Proceedings of the 34th International Conference on Machine Learning*, 70, 933–941.

Dehghani, M., Gouws, S., Vinyals, O., Uszkoreit, J., & Kaiser, L. (2019). *Universal transformers. 7th International Conference on Learning Representations, ICLR 2019,* 1–23.

Devlin, J., Chang, M.-W., Lee, K., & Toutanova, K. (2019). BERT: pre-training of deep bidirectional transformers for language understanding. *Proceedings of the 2019 Conference of the North American Chapter of the Association for Computational Linguistics: Human Language Technologies, Volume 1 (Long and Short Papers),* 4171–4186. https://doi.org/10.18653/v1/N19-1423

Devlin, J., Zbib, R., Huang, Z., Lamar, T., Schwartz, R., & Makhoul, J. (2014). Fast and robust neural network joint models for statistical machine translation. *Proceedings of the 52nd Annual Meeting of the Association for Computational Linguistics,* 1370–1380. https://doi.org/10.3115/v1/p14-1129

Doddington, G. (2002). Automatic evaluation of machine translation quality using n-gram co-occurrence statistics. *HLT '02: Proceedings of the Second International Conference on Human Language Technology Research,* 138–145. https://doi.org/10.3115/1289189.1289273

Duchi, J., Hazan, E., & Singer, Y. (2011). Adaptive subgradient methods for online learning and stochastic optimization. *Journal of Machine Learning Research,* 12, 2121–2159.

Edunov, S., Ott, M., Auli, M., & Grangier, D. (2020). Understanding back-translation at scale. *Proceedings of the 2018 Conference on Empirical Methods in Natural Language Processing, EMNLP 2018,* 489–500. https://doi.org/10.18653/v1/d18-1045

Friedman, J. H. (2002). Stochastic gradient boosting. *Computational Statistics and Data Analysis,* 38(4), 367–378.

Gehring, J., Auli, M., Grangier, D., Yarats, D., & Dauphin, Y. N. (2017). *Convolutional sequence to sequence learning. Proceedings of the 34th International Conference on Machine Learning,* 1243–1252. https://arxiv.org/abs/1705.03122

Graves, A. Mohamed, A., & Hinton, G. (2013). Speech recognition with deep recurrent neural networks. *2013 IEEE International Conference on Acoustics, Speech and Signal Processing,* 6645–6649. https://doi.org/10.1109/ICASSP.2013.6638947

Graves, A. (2012). Sequence transduction with recurrent neural networks. *ArXiv E-Prints,* arXiv:1211.3711. https://arxiv.org/abs/1211.3711

Gu, J., Lu, Z., Li, H., & Li, V. O. K. (2016). Incorporating copying mechanism in sequence-to-sequence learning. *54th Annual Meeting of the Association for Computational Linguistics, ACL 2016- Long Papers,* 3, 1631–1640. https://doi.org/10.18653/v1/p16-1154

Gu, J., Wang, Y., Chen, Y., Cho, K., & Li, V. O. K. (2018). Meta-learning for low-resource neural machine translation. *Proceedings of the 2018 Conference on Empirical Methods in Natural Language Processing, EMNLP 2018,* 3622–3631. https://doi.org/10.18653/v1/D18-1398

Gulcehre, C., Ahn, S., Nallapati, R., Zhou, B., & Bengio, Y. (2016). Pointing the unknown words. *Proceedings of the 54th Annual Meeting of the Association for Computational Linguistics,* 140–149.

Gulcehre, C., Firat, O., Xu, K., Cho, K., Barrault, L., Lin, H.-C., Bougares, F., Schwenk, H., & Bengio, Y. (2015). *On using monolingual corpora in neural machine translation. ArXiv E-Prints,* arXiv:1503.03535. https://arxiv.org/abs/1503.03535

Guzmán, F., Joty, S., Màrquez, L., & Nakov, P. (2017). Machine translation evaluation with neural networks. *Computer Speech and Language*, *45*, 180–200.
https://doi.org/10.1016/j.csl.2016.12.005

Hochreiter, S. (1998). The vanishing gradient problem during learning recurrent neural nets and problem solutions. *International Journal of Uncertainty, Fuzziness
and Knowledge-Based Systems*, *6*(2), 107–116. https://doi.org/10.1142/
S0218488598000094

Hochreiter, S., & Schmidhuber, J. (1997). Long short-term memory. *Neural
Computation*, *9*(8), 1735–1780.

Hutchins, J. (1997). Looking back to 1952: the first MT conference. *Proceedings
of the the Conference on Theoretical and Methodological Issues in Machine
Translation of Natural Languages*, 19–30.

Isozaki, H., Hirao, T., Duh, K., Sudoh, K., & Tsukada, H. (2010). Automatic evaluation of translation quality for distant language pairs. *EMNLP 2010- Conference
on Empirical Methods in Natural Language Processing, Proceedings of the
Conference, October*, 944–952.

Jean, S., Cho, K., Memisevic, R., & Bengio, Y. (2015a). On using very large target vocabulary for neural machine translation. *Proceedings of the 53rd
Annual Meeting of the Association for Computational Linguistics and the 7th
International Joint Conference on Natural Language Processing*, 1–10. https://
doi.org/10.3115/v1/P15-1001

Jean, S., Firat, O., Cho, K., Memisevic, R., & Bengio, Y. (2015b). Montreal neural
machine translation systems for WMT'15. *Proceedings of the Tenth Workshop
on Statistical Machine Translation*, 134–140. https://doi.org/10.18653/v1/
W15-3014

Joshi, M., Chen, D., Liu, Y., Weld, D. S., Zettlemoyer, L., & Levy, O. (2020). SpanBERT:
improving pre-training by representing and predicting spans. *ArXiv E-Prints*,
arXiv:1907.1052.

Kaiser, Ł., & Bengio, S. (2016). Can active memory replace attention? *Proceedings of
the 30th International Conference on Neural Information Processing Systems*,
3781–3789.

Kaiser, Ł., & Sutskever, I. (2015). Neural GPUs learn algorithms. *ArXiv E-Prints*,
arXiv:1511.08228.

Kalchbrenner, N., & Blunsom, P. (2013a). Recurrent continuous translation
models. *Proceedings of the 2013 Conference on Empirical Methods in
Natural Language Processing*, 1700–1709. https://doi.org/10.1146/annurev.
neuro.26.041002.131047

Kalchbrenner, N., & Blunsom, P. (2013b). Recurrent convolutional neural networks
for discourse compositionality. *Proceedings of the Workshop on Continuous
Vector Space Models and Their Compositionality*, 119–126.

Kalchbrenner, N., Espeholt, L., Simonyan, K., Oord, A. van den, Graves, A., &
Kavukcuoglu, K. (2016). Neural machine translation in linear time. *ArXiv
E-Prints*, arXiv:1610.10099.

Kingma, D. P., & Ba, J. L. (2014). Adam: a method for stochastic optimization.
Arxiv:1412.6980. https://arxiv.org/abs/1412.6980

Koehn, P. (2010). *Statistical Machine Translation* (1st ed.). Cambridge University
Press, New York.

Koehn, P., & Knowles, R. (2017). Six challenges for neural machine translation. *Proceedings of the First Workshop on Neural Machine Translation*, 28–39. https://doi.org/10.18653/v1/w17-3204

Koehn, P., Och, F. J., & Marcu, D. (2003). Statistical phrase-based translation. *Proceedings of the 2003 Conference of the North American Chapter of the Association for Computational Linguistics on Human Language Technology - Volume 1*, 48–54. https://doi.org/10.3115/1073445.1073462

Krizhevsky, A., Sutskever, I., & Hinton, G. E. (2012). ImageNet classification with deep convolutional neural networks. *Advances in Neural Information Processing Systems*, 1097–1105.

Lample, G., Conneau, A., Denoyer, L., & Ranzato, M. (2017). Unsupervised machine translation using monolingual corpora only. *ArXiv E-Prints*, arXiv:1711.00043.

Lample, G., Ott, M., Conneau, A., Denoyer, L., & Ranzato, M. (2020). Phrase-based & neural unsupervised machine translation. *Proceedings of the 2018 Conference on Empirical Methods in Natural Language Processing, EMNLP 2018*. https://doi.org/10.18653/v1/d18-1549

Lan, Z., Chen, M., Goodman, S., Gimpel, K., Sharma, P., & Soricut, R. (2020). ALBERT : a lite BERT for self-supervised learning of language representations. *International Conference on Language Resources and Evaluation*, 1–17.

Lavie, A., & Denkowski, M. J. (2009). The METEOR metric for automatic evaluation of machine translation. *Machine Translation*, 23(2–3), 105–115. https://doi.org/10.1007/s10590-009-9059-4

Lee, J., Cho, K., & Hofmann, T. (2017). Fully character-level neural machine translation without explicit segmentation. *Transactions of the Association for Computational Linguistics*, 5, 365–378. https://arxiv.org/abs/1610.03017

Ling, W., Trancoso, I., Dyer, C., & Black, A. W. (2016). Character-based neural machine translation. *Proceedings of the 54th Annual Meeting of the Association for Computational Linguistics*, 357–361. https://doi.org/10.18653/v1/P16-2058

Liu, Y., Ott, M., Goyal, N., Du, J., Joshi, M., Chen, D., Levy, O., Lewis, M., Zettlemoyer, L., & Stoyanov, V. (2019). RoBERTa: a robustly optimized BERT pretraining approach. *ArXiv E-Prints*, arXiv:1907.1169.

Luong, M.-T. (2017). *Neural Machine Translation*. Stanford University, Stanford, CA.

Luong, M.-T., & Manning, C. D. (2016). Achieving open vocabulary neural machine translation with hybrid word-character models. *54th Annual Meeting of the Association for Computational Linguistics, ACL 2016- Long Papers*, 2, 1054–1063. https://doi.org/10.18653/v1/p16-1100

Luong, M.-T., Pham, H., & Manning, C. D. (2015a). Effective approaches to attention-based neural machine translation. *Proceedings of the 2015 Conference on Empirical Methods in Natural Language Processing*, 1412–1421.

Luong, M.-T., Sutskever, I., Le, Q. V, Vinyals, O., & Zaremba, W. (2015b). Addressing the rare word problem in neural machine translation. *Proceedings of the 53rd Annual Meeting of the Association for Computational Linguistics and the 7th International Joint Conference on Natural Language Processing*, 11–19.

Maas, A. L., Hannun, A. Y., & Ng, A. Y. (2013). Rectifier nonlinearities improve neural network acoustic models. *Proceedings of the 30th International Conference on Machine Learning*, 28.

Mikolov, T., Corrado, G., Chen, K., & Dean, J. (2013a). Efficient estimation of word representations in vector space. *1st International Conference on Learning Representations (ICLR) 2013*. https://arxiv.org/abs/1301.3781

Mikolov, T., Sutskever, I., Chen, K., Corrado, G., & Dean, J. (2013b). Distributed representations of words and phrases and their compositionality. *Proceedings of the Advances Neural Information Processing Systems*, 3111–3119.

Neubig, G., & Hu, J. (2018). Rapid adaptation of neural machine translation to new languages. *Proceedings of the 2018 Conference on Empirical Methods in Natural Language Processing*, 875–880. https://doi.org/10.18653/v1/D18-1103

Nguyen, T. Q., & Chiang, D. (2017). Transfer learning across low-resource, related languages for neural machine translation. *Proceedings of the Eighth International Joint Conference on Natural Language Processing (Volume 2: Short Papers)*, 296–301.

Nguyen, T., & Chiang, D. (2018). Improving lexical choice in neural machine translation. *Proceedings of the 2018 Conference of the North {A}merican Chapter of the Association for Computational Linguistics: Human Language Technologies, Volume 1 (Long Papers)*, 334–343. https://doi.org/10.18653/v1/N18-1031

tling, R., & Tiedemann, J. (2017). Neural machine translation for low-resource languages. *ArXiv E-Prints*, arXiv:1708.05729.

Papineni, K., Roukos, S., Ward, T., & Zhu, W.-J. (2002). Bleu: a method for automatic evaluation of machine translation. *Proceedings of the 40th Annual Meeting of the Association for Computational Linguistics*, 311–318. https://doi.org/10.3115/1073083.1073135

Pascanu, R., Mikolov, T., & Bengio, Y. (2012). *Understanding the exploding ggradient problem*. CoRR, abs/1211.5063.

Pascanu, R., Mikolov, T., & Bengio, Y. (2013). On the difficulty of training recurrent neural networks. *30th International Conference on Machine Learning, ICML 2013*, PART 3, 2347–2355.

Pennington, J., Socher, R., & Manning, C. D. (2014). GloVe : global vectors for word representation. *Proceedings of the 2014 Conference on Empirical Methods in Natural Language Processing (EMNLP)*, 1532–1543.

Peters, M. E., Neumann, M., Iyyer, M., Gardner, M., Clark, C., Lee, K., & Zettlemoyer, L. (2018). Deep contextualized word representations. *Proceedings of the 2018 Conference of the North American Chapter of the Association for Computational Linguistics: Human Language Technologies, Volume 1 (Long Papers)*, 2227–2237. https://doi.org/10.18653/v1/N18-1202

Radford, A., Narasimhan, K., Salimans, T., & Sutskever, I. (2018a). Improving Language Understanding by Generative Pre-Training. *OpenAI Preprint*.

Radford, A., Wu, J., Child, R., Luan, D., Amodei, D., & Sutskever, I. (2018b). Language Models are Unsupervised Multitask Learners. *OpenAI Preprint*.

Ren, S., Zhang, Z., Liu, S., Zhou, M., & Ma, S. (2019). Unsupervised neural machine translation with SMT as posterior regularization. *The Thirty-Third AAAI Conference on Artificial Intelligence 2019*, 241–248. https://doi.org/10.1609/aaai.v33i01.3301241

Rumelhart, D. E., Hinton, G. E., & Williams, R. J. (1986). Learning representations by back-propagating errors. *Nature, 323*, 533–536.

Schuster, M., & Paliwal, K. K. (1997). Bidirectional recurrent neural networks. *IEEE Transactions on Signal Processing, 45*(11), 2673–2681.

Schwenk, H. (2012). Continuous space translation models for phrase-based statistical machine translation. *Proceedings of COLING 2012*, 1071–1080. https://aclweb.org/anthology-new/C/C12/C12-2104.pdf

Schwenk, H., Chaudhary, V., Sun, S., Gong, H., & Guzmán, F. (2019a). WikiMatrix: mining 135M parallel sentences in 1620 language pairs from Wikipedia. CoRR, abs/1907.05791. https://arxiv.org/abs/1907.05791

Schwenk, H., Wenzek, G., Edunov, S., Grave, E., & Joulin, A. (2019b). CCMatrix: mining billions of high-quality parallel sentences on the {WEB}. *CoRR, abs/1911.04944*. https://arxiv.org/abs/1911.04944

Sennrich, R., & Haddow, B. (2016). Linguistic input features improve neural machine translation. *Proceedings of the First Conference on Machine Translation*, 83–91. https://doi.org/10.18653/v1/w16-2209

Sennrich, R., Haddow, B., & Birch, A. (2016). Improving neural machine translation models with monolingual data. *54th Annual Meeting of the Association for Computational Linguistics, ACL 2016- Long Papers, 1*, 86–96. https://doi.org/10.18653/v1/p16-1009

Sennrich, R., Haddow, B., & Birch, A. (2015). Neural machine translation of rare words with subword units. *Proceedings of the 54th Annual Meeting of the Association for Computational Linguistics*, 1715–1725. https://arxiv.org/abs/1508.07909

Sennrich, R., & Zhang, B. (2019). Revisiting low-resource neural machine translation: a case study. *Proceedings of the 57th Annual Meeting of the Association for Computational Linguistics*, 211–221. https://doi.org/10.18653/v1/P19-1021

Settles, B. (2012). Active learning. Synthesis Lectures on Artificial Intelligence and Machine Learning. Springer, Cham. https://doi.org/10.1007/978-3-031-01560-1.

Shaw, P., & Uszkoreit, J. (2018). Self-attention with relative position representations. *Proceedings of the 2018 Conference of the North American Chapter of the Association for Computational Linguistics: Human Language Technologies, Volume 2 (Short Papers)*, 464–468. https://doi.org/10.18653/v1/N18-2074

Shazeer, N., Mirhoseini, A., Maziarz, K., Davis, A., Le, Q., Hinton, G., & Dean, J. (2017). Outrageously large neural networks: The sparsely-gated mixture-of-experts layer. *ArXiv E-Prints*, 1701.06538.

So, D. R., Liang, C., & Le, Q. V. (2019). The evolved transformer. *Proceedings of the 36th International Conference on Machine Learning*, Long Beach, CA, 5877–5886.

Søgaard, A., Ruder, S., & Vulić, I. (2018). On the limitations of unsupervised bilingual dictionary induction. *Proceedings of the 56th Annual Meeting of the Association for Computational Linguistics (Volume 1: Long Papers)*, 778–788. https://doi.org/10.18653/v1/P18-1072

Sundermeyer, M., Alkhouli, T., Wuebker, J., & Ney, H. (2014). Translation modeling with bidirectional recurrent neural networks. *Proceedings of the 2014 Conference on Empirical Methods in Natural Language Processing (EMNLP)*, 14–25. https://doi.org/10.3115/v1/D14-1003

Sutskever, I., Vinyals, O., & Le, Q. V. (2014). Sequence to sequence learning with neural networks. *Advances in Neural Information Processing Systems*, 3104–3112. https://arxiv.org/abs/1409.3215

Tab-delimited Bilingual Sentence Pairs. (2021). https://www.manythings.org/anki/

Tatoeba. (2021). https://tatoeba.org/en

Van Houdt, G., Mosquera, C., & Nápoles, G. (2020). A review on the long short-term memory model. *Artificial Intelligence Review*, 53(8), 5929–5955. https://doi.org/10.1007/s10462-020-09838-1

Vaswani, A., Shazeer, N., Parmar, N., Uszkoreit, J., Jones, L., Gomez, A. N., Kaiser, L., & Polosukhin, I. (2017). Attention is all you need. *ArXiv E-Prints*, arXiv:1706.03762. https://arxiv.org/abs/1706.03762

Wang, Q., Li, B., Xiao, T., Zhu, J., Li, C., Wong, D. F., & Chao, L. S. (2019a). Learning deep transformer models for machine translation. *Proceedings of the 57th Annual Meeting of the Association for Computational Linguistics*, 1810–1822.

Wang, W., Bi, B., Yan, M., Wu, C., Bao, Z., Xia, J., Peng, L., & Si, L. (2019b). StructBERT: incorporating language structures into pre-training for deep language understanding. *ArXiv E-Prints*, arXiv:1908.04577.

Werbos, P. J. (1990). Backpropagation through time: what it does and how to do it. *Proceedings of the IEEE*, 78(10), 1550–1560. https://doi.org/10.1109/5.58337

Wu, Y., Schuster, M., Chen, Z., Le, Q. V, Norouzi, M., Macherey, W., Krikun, M., Cao, Y., Gao, Q., Macherey, K., Klingner, J., Shah, A., Johnson, M., Liu, X., Kaiser, Ł., Gouws, S., Kato, Y., Kudo, T., Kazawa, H., ... Dean, J. (2016). Google's neural machine translation system: bridging the gap between human and machine translation. *ArXiv E-Prints*, arXiv:1609.08144. https://arxiv.org/abs/1609.08144

Yang, Z., Dai, Z., Yang, Y., Carbonell, J., Salakhutdinov, R., & Le, Q. V. (2019). XLNet : generalized autoregressive pretraining for language understanding. *33rd Conference on Neural Information Processing Systems* (NeurIPS 2019), 1–18.

Zeiler, M. D., & Fergus, R. (2014). Visualizing and understanding convolutional networks. Lecture Notes in Computer Science (Including Subseries Lecture Notes in Artificial Intelligence and Lecture Notes in Bioinformatics), 8689 LNCS(PART 1), 818–833. https://doi.org/10.1007/978-3-319-10590-1_53

Zhou, J., Cao, Y., Wang, X., Li, P., & Xu, W. (2016). Deep recurrent models with fast-forward connections for neural machine translation. *Transactions of the Association for Computational Linguistics*, 4, 371–383. https://doi.org/10.1162/tacl_a_00105

Zoph, B., Yuret, D., May, J., & Knight, K. (2016). Transfer learning for low-resource neural machine translation. *EMNLP 2016- Conference on Empirical Methods in Natural Language Processing, Proceedings*. https://doi.org/10.18653/v1/d16-1163

Chapter 5

Evolution of question-answering system from information retrieval

A scientific time travel for Bangla

Arijit Das and Diganta Saha
Jadavpur University

5.1 INTRODUCTION

It was 1960 when a group of scientists involved in the defense projects of the federal government of the USA felt the need to connect their computers to share the jobs done and avoid redoing the same job by different scientists. The first network connection is called ARPANET. With time, networks in universities, research centers, and industries got connected with the help of funding from the National Science Foundation and many other major private players. Internetworking protocols like TCP/IP were published in 1974. The domain and scope of "search" were broadened with the internet, from single personal computers to all networked computers.

Around 1985, the internet was widespread in Europe. Berners Lee in the CERN lab was frustrated with the difficulties faced in locating or searching the information stored on the different computers. In 1990, he and his colleague Robert Cailliau introduced the concepts of hypertext, http, and the World Wide Web or www. They proposed to structure the document or information in hypertext and access it via a Uniform Resource Locator (URL) with the help of the Domain Name System (DNS). They became instrumental in developing web browser software to access any HTML document hosted by the web server in the client-server architecture. Now the specialization of "search" started to intensify toward "web search."

The invention of www started a race among organizations to bring their presence to the internet through websites. A huge boom of data or information over the web made way for web search engines.

Companies like Yahoo, Google, and Microsoft launched their search engines. They were primarily working on syntax matching on the repository made by their web crawlers, and then returned the list of URLs rankwise in response to the search. The list was made based on a matching index. Later, the search engines started improving the ranking list by applying filters like user preference, location, history, etc.

DOI: 10.1201/9781003244332-5

Some of the milestones in the timeline of the search industry are as follows:

- 1995 – Yahoo search was launched.
- 1996 – Seznam was incorporated. It is dominant in the Czech Republic.
- 1997 – Google was launched.

Yandex was launched. Yandex is dominant in Russia.

- 1998 – Microsoft launched MSN search.
- 1999 – Naver was incorporated. It is dominant in South Korea.
- 2000 – Baidu was incorporated. It is dominant in China.
- 2008 – DuckDuckGo was introduced. It has become famous for anonymous searching.
- 2013 – Qwant was launched. It is dominant in France.

Since the inception of Google in 1998, industry has started to feel how search has become a dominant technology due to the rapid growth of the internet and data. Undoubtedly, Google has become the overall global market leader in web search. The popularity of search engines like Seznam, Yandex, Naver, Baidu, and Qwant in the Czech Republic, Russia, South Korea, China, and France was greater than that of Google for their accurate results in the Czech, Russian, Korean, Chinese, and French languages, respectively.

Researchers, Scientists, and Engineers felt a new dimension of search was needed to understand the meaning of language and the context of search terms from the user's perspective. From 2004 onwards, several papers on semantic search were published that promised to produce precise answers to user queries by taking advantage of the availability of explicit semantics of information.

5.2 THE MEANING AND VARIOUS WAYS OF RESEARCH DONE IN THE FIELD OF SEMANTICS

The word "Semantics" is a noun. It is the branch of linguistics where studies regarding the meaning of texts are performed. The adjective is "Semantic." The word originated from the Greek word "Sema." The word was included in the English vocabulary in the mid-17th century. The morphology is shown in Figure 5.1.

From 1920 onwards, use of the word "semantics" was noticed in English literature and newspapers, and from 1960 onwards, the frequency of use of the word has increased (Data collected from Google N-gram viewer) to a considerable amount.

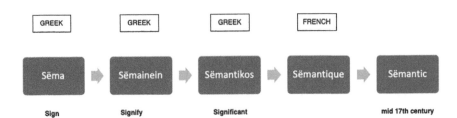

Figure 5.1 Morphological change of the word semantics.

In the field of Computer Science, research around semantics is multidimensional. Analogously, the word "Informatics" is used to organize the information of a specific field. The concatenation of the term "informatics" with different fields signifies different fields of study or research, such as Bio-informatics, Geo-informatics, Health-informatics, etc. Similarly, the term "semantic" used with the field name represents different fields of study or research. Some fields use "semantic" in nomenclature, as represented in the list with their objectives.

- **Semantic web** – This is the field where scientists work toward organizing internet data in a machine-readable format.
- **Semantic music** – In this field, scientists work to extract the meaning of music, irrespective of language, using all the musical attributes. Relating two songs of different languages or different melodies is the objective of this field.
- **Semantic video** – Processing of video, segmentation, finding the inner meaning of a video clip, and labeling are the major objectives of the field. The information flows from one frame to the next, and particular messages or information in one frame are handled here. The major challenges are classifying videos according to the meaning and similarity of two clips and finding specific search objects in the video.
- **Semantic image** – This is a specialized field of image processing where the system extracts the meaning or context of the image. This field's main objectives are automatic image description generation, image classification, and labeling.
- **Semantic text** – In this field of research, the system extracts or understands the meaning of the text. Similarity measurements between two words or two sentences or two paragraphs, or two corpora are the generic challenges of this field of research. Instead of keyword matching, semantic text objectifies the process of extracting and labeling the meaning of the text.
- **Semantic graph** – Semantic graph is a tool to represent the semantic relationship between objects (text/image/video/music). It forms a network and a directed or undirected graph where nodes or vertices

represent the concept, and the edges represent the relations among those concepts.

- **Semantic search** – As the name suggests, Semantic Search is the process of searching semantically or contextually. The semantic search domain is huge and can range into different mediums such as text, image, audio, video, web links, etc. Query or question processing is common for all types of search – text, image, audio, or video. The meaning of the query is understood in the query or question processing phase. In the case of text, an effort is made to answer the query or questions instead of giving a list of URLs based on keyword matching.

Among all these, semantic search and semantic web have attracted the most attention among the researchers. My work is in the text field of semantic search, with an interest in the Bangla language. So, the following parts of the survey have been done accordingly, synchronizing my research's interest, scope, and domain.

5.3 SEMANTIC TEXT RETRIEVAL: AUTOMATIC QUESTION-ANSWERING SYSTEM OR QUERY SYSTEM

The system implementing semantic text search aims to understand the context and meaning of the question and then give a meaningful answer from the repository with the help of a knowledge base and algorithms. The query or question in the search system can be given as a voice command or text command. Similarly, the output can also be in the form of voice or text. In the case of voice search, generally, voice-to-text software is used to convert the input to text, and text-to-voice software is used to convert the output to voice. Accent, pronunciation, vocal tone, etc. play crucial roles on top of core search algorithms in the case of voice search. In my work, we have concentrated on core textual search algorithms. So, both input and output are considered to be text.

On the other hand, the question-answering (QA) system is a special kind of search where the realization of semantic search can be found. Depending on various parameters, the research on QA systems is diversified in different dimensions.

The repository used for the QA system may be structured, semi-structured, or unstructured. A relational database is an example of a structured repository. For such a QA system, the question asked in natural language is converted to SQL or Structured Query Language. An example of a semi-structured repository is XML or eXtensible Markup Language. In the case of the semantic web, Resource Data Framework or RDF is used as a data model, which is also a semi-structured repository. DBpedia is one of the famous examples of semi-structured repositories which extract metadata from Wikipedia. Briefly, a semi-structured repository is a database with a less strict structure definition than a structured database. Even MongoDB or other NoSQL databases can be fitted into semi-structured repository classes. Database-specific languages are used for querying the semi-structured repository. SPARQL is used for RDF, XQuery, or XML-QL is used for XML. In the case of the semi-structured repository, the question in natural language is converted to the requisite query language (like SPARQL), depending on the repository. The unstructured repository is the file containing free text. The collection of text or word files is an example of an unstructured repository. Where structured and semi-structured databases are the synthetic repositories produced by some processing, the unstructured repository is the simple raw file system. Statistically, unstructured free text is the most available digital content in any low-resource language concerning the structured or semi-structured repository.

Depending on the type of response, QA systems can be of three types. First, multiple options are given as a response by the system. IVRS and MCQ are examples of this category. In the second category, an objective answer to the question is expected. The Factoid QA system is an example of this category. Third, a descriptive answer to the question is expected. The narrative QA system is an example of this category. A text generation technique is required in this third category.

5.4 STATE-OF-THE-ART PERFORMANCE

The history of the QA system dates back at least 60 years. BASEBALL [1] is considered one of the first computer-based QA systems. The program "BASEBALL" was developed at the Lincoln Laboratory of Massachusetts Institute of Technology, USA, receiving grants from the US army, air force, and navy. It received questions in ordinary English, processed a structured list database, and returned the answer. After reading the question from the punch cards, BASEBALL looked up dictionaries for synonyms, idioms, phrases, etc. The English dictionary was stored as an attribute-value pair. Then the requested data or answer was extracted from the matching record in the database. Primarily, it was designed to collect data regarding baseball tournaments in the USA.

In 1973, a similar system named LUNAR [2] was developed. It was a QA system accepting questions in ordinary English and giving answers related to the geological data collected by the US space program on the Moon. By the time project [1] was being developed, Joseph Weizenbaum had developed ELIZA [3], which was one of the earliest natural language processing (NLP) programs and a chatbot. Its most famous script is DOCTOR, a simulated psychotherapist. It was used to communicate with patients and is believed to affect them positively. It was also helpful for the doctors to read the minds of patients. ELIZA worked based on pattern matching and NLP. NLP programming techniques were also used in references [1] and [2]. It did not have a framework. For every pattern of questions, a script was written to generate a "parrot-like" response. With such early research, it became prominent that

    a. improvement of the QA system is dependent mainly on NLP.
    b. language-specific resources or techniques are required to build such QA systems in a specific language.
    c. search engines and QA systems are closely related in terms of technology. Some of the techniques can be used interchangeably.
    d. chatbot is one of the major applications of the QA system.

The year 2004 was one of the greatest milestones in QA research. In 2004, IBM Research manager Charles Lickel watched a restaurant's famous quiz show, "Jeopardy," on television. He coined the idea to build a system that could be a contestant in "Jeopardy." IBM initiated the "DeepQA" project and developed the system named after IBM's first CEO and founder Thomas J. Watson. Watson [4] could understand the questions in natural language and answer them by processing its repository. In 2011, Watson became the champion of Jeopardy. In 2013, IBM started the commercial application of Watson. The success of Watson was multidimensional.

5.5 LATEST RESEARCH WORKS ON QUESTION-ANSWERING SYSTEMS IN RELATED MAJOR GLOBAL LANGUAGES

Chatbots and virtual assistants have started dominating user interfaces. Various industries, especially service-based ones, have started to employ chatbots to reduce pressure on customer care departments. Virtual assistants like Siri, Google Assistant, Cortana, and Alexa start their journey as integral parts of different commercial operating systems.

Researchers, scientists, and academicians in NLP and Information Retrieval fields have started working to build QA systems in various regional languages all over the globe using the latest machine learning and deep learning techniques. Machine translation cannot answer when the question and repository are both in the same native regional natural language. Although primarily the QA system was built in English, more than 620 crores of the population do not speak English, out of the total global population of 750 crores. In multilingual countries and rural areas, people can communicate only in their mother tongues. Thus, the huge demand for QA systems that can understand questions in popular natural languages and answer processing repositories boosted research for dynamic QA systems in different languages across the globe.

Historically, foreigners of diverse origins and cultures have come to India from time to time to rule, study, or live as migrants. Similarly, ambassadors and scholars from the ancient world were exchanged. Indian languages and cultures are inflected with those foreign languages. Thus, those languages have a relationship to Indian ethnicity. The latest QA systems in various related global languages (which have a connection with Indian ethnicity) and their research features are discussed in the following few paragraphs.

Gao and his group [5] achieved 100% accuracy in the supervised method of the QA system in the movie-related English questions. They used 2,60,000 training questions to test 29,000 test questions. Their approach is termed Knowledge Authoring Learning Machine (KALM), which was claimed to perform better than the machine learning approach.

Liu et al. [6] carried out English question classification to classify the questions into one of the predefined categories with semantic feature extraction and lexical feature extraction. The approach achieved 90.4% fine-grain accuracy and 96% coarse-grain accuracy over the UIUC dataset.

Radoev and the group [7] built the Language Adaptive Method for question-Answering (LAMA) system for answering natural language questions in English and French over a semi-structured DBPedia database. They defined lexical and syntactic patterns to convert the question into SPARQL queries.

In his thesis [8], Kyle Moore approached using online forums as an unstructured knowledge base to answer the questions in English natural language. Data were collected mainly from Reddit and Stack Exchange. The summarization method was used to generate the candidates of answers and then combine them.

Dennis and the group [9] translated natural language queries in English, German, French, Italian, and Spanish to SPARQL queries to be executed in a semi-structured database or repository such as DBPedia, DBLP, etc.

Liu et al. [10] developed a deep reinforcement learning-based approach to refine the questions in English. Refined questions are generated from erroneous or ambiguous questions through sequential steps of reinforcement learning. The experiment showed substantial improvement in the retrieval of accurate answers.

Readily available questions, repositories, and human-evaluated answers help test the algorithm in less time and compare the result with benchmark values. Stanford Question-Answering Dataset (SQuAD), containing one million questions generated by volunteers [11] working in the crowdsource model, was made publicly available in 2016. This helped the researchers quickly test and compare their models, algorithms, and results over readily available English comprehension, readymade questions, and human-given answers.

SQuAD version 2.0 was released in 2018 [12]. It included 50,000 unanswerable questions with no answers in the repository to improve the reliability of the analysis of the models.

A detailed list of research carried out on the SQuAD is given in [13]. Here, the most notable and useful projects with results are discussed. In 2017, researchers from Google applied Bidirectional Encoder Representations from Transformers (BERT) [14] on the SQuAD to retrieve answers more accurately and efficiently. Lan et al. [15] proposed "A Lite BERT" or "ALBERT" to develop QA systems. It gave a 92.215 F1 score. Later, after multiple attempts, Zhang and the group [16] designed a retro reader on top of the ALBERT, which gave a 92.978 F1 score.

In Table 5.1, major languages that have connections with Indian ethnicity are listed from different continents. Next, promising research on the QA system in recent days in those major foreign languages is discussed. Challenges due to the unavailability of large-scale datasets (such as SQuAD) for experiments in non-English QA systems are also described.

In Spanish, examples of current research on the QA system are [17,18]. Segura et al. [17] developed a chatbot named "Chatbol" to answer questions in the natural Spanish language regarding the famous Spanish football tournament "La Liga." "Chatbol" achieved 68% accuracy for answering casual chatting and 72% accuracy for answering questions. Carrino and the group depicted the unavailability of large-scale datasets for experiments on the QA system in a non-English language. Therefore, they translated the SQuAD dataset into Spanish.

In the French language, [19] and [20] are examples of recent research trends in the QA system. Radoev and the group [19] concentrated on transforming the question in French natural language into a SPARQL query for the DBPedia database. This is claimed as the first work in the French QA system. Keraron et al. [20] discussed the unavailability of the large-scale dataset for experiments on the QA system in the French language. Thus, they tried to build a QA dataset in French, such as SQuAD. French

Table 5.1 Major languages in different continents

Continents	Major foreign languages having connections with Indian ethnicity
Europe	English [5–16], Spanish [17,18], French [19,20], German [21], Portuguese [22–24], Italian [25–28], Greek, Russian
Asia	Indonesian [29,30], Japanese [31–33], Chinese [34–44], Tibetan [45–47], Arabic [48–54], Persian [55–57], Hebrew
North America	English [5–16], Spanish [17,18], French [19,20]
South America	Spanish [17,18], Portuguese [22–24]
Australia or Occania	English [5–16]
Africa	Somali, Niger
Antarctica	Russian

Wikipedia, question crowdsourcing, and machine translation were the main strategies to present such a dataset.

In the German language, Falke [21] used a knowledge graph to answer questions in the German natural language, which is the current work in the German QA system. The domain chosen for the system was question-answering German grammar.

In the Portuguese language, [22–24] is the recent research on the QA system. In his doctoral work, Rodrigues [22] developed a system called "RAPPORT" to answer questions in the Portuguese natural language. A collection of Portuguese news items called "CHAVE," having a size of 108 MB, is used as the repository. It recorded 47% accuracy initially for the QA system over "CHAVE." Rodrigues and Gomes [23] developed a QA system for answering questions in Portuguese natural language by extracting triples (subject, object, and predicate) from the sentence and achieved 42% accuracy for the "CHAVE" repository. Araujo and the group [24] concentrated on developing a QA system for the Brazilian Portuguese language. They transformed the question in natural language Portuguese into a SPARQL query to address the DBPedia repository.

In the Italian language, [25–28] is recent research on the QA system. Damiano [25] and the group used deep learning and a rule-based approach for answering questions in the Italian natural language in the "Cultural Heritage" closed domain. Croce and the group [26] used 60,000 question-answer pairs from the Italian Wikipedia to train the system using deep learning. Polignano et al. [27] did the QA experiment on Tweet data in Italian. They modified the classical BERT model for Italian language and social data, called it AlBERTo, and achieved a 94.03 F1 score as the best case. Cuteri and the group [28] developed the QA system in the closed domain "Cultural Heritage." The question in natural language is transformed into a query by rule-based syntactic classification.

In the Indonesian language, [29] and [30] are the recent research on the QA system. Gunawan and the group [29] used NLP techniques and pattern matching to answer arithmetic problems expressed in the natural Indonesian language. The accuracy ranged from 80% to 100%, depending on the complexity. The average response time was 1.12 minutes. Putra and the group [30] used Lucerne's score and Wijono's algorithm to find the similarity between the question or query and the predicted answer text.

The QA system's recent research in the Japanese language [31–33]. Day et al. [31] developed a QA system in the Japanese language for the university entrance examination using NLP and ML techniques. The question in Japanese is translated to English using Google Translate, and then the answer is retrieved from the available sources in English. The retrieved answer in English is then translated back to Japanese. The main drawback of such a translation technique is that the system fails to answer if the answer is available only in Japanese and not in English. Sakamoto et al.

[32] developed a QA system in the Japanese language in both MCQ and descriptive format for the university exam for the History subject. World history textbooks and the World history glossary are used as repositories. Kato et al. [33] compared LSTM and BM25 at the algorithmic level for the efficiency and effectiveness of the QA system in the Japanese language.

Chinese is an umbrella language with dialects such as Mandarin, Cantonese, Hunanese, etc. In the Chinese language, [34–44] is the recent research on the QA system. Liu et al. [34] described the unavailability of a large corpus in the Chinese language. A corpus is created as the Chinese QA systems dataset with question-answer pairs. The dataset was filtered with Wasserstein Distance to remove irrelevant pairs; ultimately, 2,60,068 questions are present in the dataset. A baseline approach for future research in this dataset is also provided. Chen and the group [35] used a knowledge graph to build QA systems in classical Chinese poetry. Zhou and the group [36] developed an intelligent QA system on a Chinese medical text corpus based on a score fusion algorithm and achieved 78.49% accuracy for answer retrieval. The group has claimed to answer more complex questions concerning other QA systems, and thus they called the system an intelligent QA system. Tian and the group [37] developed a doctor recommendation system based on online Chinese questions. Liu and the group [38] applied a convolutional neural network called attention-based Bi-GRU CNN for Chinese question classification. Tsai [39] made a study in his postgraduate work to apply techniques used in Chinese QA to other languages. Guo, Na, and Li [40] developed a resource containing an annotated, classified corpus of health questions in Chinese. The corpus contained 5,000 questions in Chinese from 29 categories.

Feng and the group [41] designed a Chinese QA system to get the actual answer from the repository. FastText model was used for question classification, and the Word2Vec model was used to retrieve the semantically related answer to the question. Liang and the group [42] used a shared convolutional network (SH-CNN) to measure semantic similarity between short-length questions and their probable answers, extracting features like Term Frequency, Inverse Document Frequency, etc. 90.70% accuracy is achieved for retrieving answers to short-length questions. Liu et al. [43] applied LSTM and CNN to understand the question and map the best answer for a question in the Chinese language and the medical domain. Character embedding is used to enhance the accuracy, which is claimed to be 80% in the best case. Another recent work on the Chinese QA system in the medical domain is done by Zhang et al. [44]. A series of simple and hybrid models of neural networks are tested on the cMedQA dataset, and 87.30% accuracy is achieved in the best case.

In the Tibetan language, [45–47] is the recent research on the QA system. Sun and the group described the problem caused by the unavailability of large-scale corpora in the Tibetan language [45] and proposed a corpus generation model for the Tibetan QA system using quasi-RNN and Reinforcement Learning. Sun and Xia [46] proposed a hybrid neural network model for Tibetan QA. An LSTM-based deep learning technique is used to extract features from small-scale Tibetan corpora. 63.1% accuracy is achieved in the best case. Zhu and the group [47] developed a deep learning-based bilingual Chinese-Tibetan QA system to address the requirements of people of an ethnic minority. Tibetan machine translation, classification, and CNN are used to carry out this work.

In Arabic, [48–54] is the current QA systems research. Mozannar and the group [48] described the problem caused by the unavailability of a large-scale QA dataset in the Arabic language. To address the problem, Arabic Reading Comprehension Dataset (ARCD), which has 1395 questions, is created from Arabic Wikipedia, and machine translation of part of SQuAD (Arabic SQuAD) is done. A document retriever using hierarchical TF-IDF and BERT is applied to retrieve the accurate answer, and 61.3 F1 scores are achieved. Hasan and the group [49] applied Support Vector Machine (SVM) and pattern matching for Arabic question classification. Approximately 200 Arabic questions are classified from Islamic Hadith into three classes – "Who," "Where", and "What" and 88.39%, 87.66%, and 87.93% F1 scores are achieved, respectively. Dardour and the group [50] used transducers and dictionaries to disambiguate various terms in the Arabic medical QA system. Muttaleb and the group [51] did an interesting survey on different aspects of the Arabic QA system. Major challenges for the QA system in Arabic are described first. Then, previous work in Arabic on question classification, their rule-based approaches, and the problems of such approaches are discussed. Bouziane and the group [52] developed an Arabic domain-independent QA system over a semi-structured dataset. The question in natural Arabic is transformed into a SPARQL query by a series of processing steps such as Tokenization, POS tagging, keyword extraction, predicate definition, etc. The query is executed to retrieve the linked semi-structured data. F1 score is claimed to improve from 70.5 to 95 after Arabic stopword removal. Al-Shenak and the group [53] used classification techniques to retrieve the relevant paragraph with answers to the Arabic question in natural language. SVM, Latent Semantic Index (LSI), and Single Value Decomposition (SVD) are used to classify the question into one of the ten classes and retrieve the accurate paragraph containing the answer with 98% accuracy. Ismail and Homsi [54] did a beautiful job creating a dataset for the Arabic "Why" QA system consisting of 3,205 "Why" QA pairs crafted from various Arabic websites. It is described that the answers to "Why" questions are generally descriptive or not factoid. Creating such a dataset will certainly help future research in the Arabic QA system.

In the Persian language, [55–57] is the current QA system research. The authors [55] developed the Persian question classification technique using machine learning methods. Feature extraction using clustering, RNN, and feed-forward networks; conversion of each question to a feature vector using Word2Vec and TF-IDF approach; and classification of the vectors or questions using SVM and neural network classifiers are the technical methods used by them. An accuracy of 76.6% is achieved in the best-case scenario regarding question classification. Boreshban and the group [56] first described the problem caused by the unavailability of a large-scale corpus for the QA system in the Persian language. Then, the development of a religious corpus for the Persian QA system, called "Rasayel&Massayel" consisting of 2,051 factoid and 2,118 non-factoid Persian questions, is discussed. Boreshban and Mirroshandel [57] developed a Persian QA system for non-factoid questions in the religious domain. After processing the questions in the Persian natural language, documents having the prospective answers are retrieved, and they are ranked. Re-ranking is done based on machine learning techniques applied to the past results of correctness. The system achieved 81.29% accuracy.

In the Russian language, [58–63] is the current research on the QA system. Nikolaev and Malafeev [58] built an automatic classifier of questions in the Russian language. Word order, pronouns, and interrogative particles are used as features to classify the questions by applying machine learning techniques (classifiers) – SVM, logistic regression, and Naïve Bayes. Fine-grain accuracy is 65.3%, and coarse-grain accuracy is 68.7%. Soboleva and Vorontsov [59] designed a Russian QA system for the corpus formed by text from Russian Wikipedia. The system involved three consecutive stages – (1) pointing to candidate documents having the prospective answer to the question, (2) ranking the sentences of the selected document according to the descending probability of having the correct answer to the question, and (3) finding the exact phrase to answer the question accurately. Almost 100% accuracy was achieved in the first stage, 97% accuracy was achieved to rank the accurate sentences within the top three results in stage 2, and 50% accuracy was achieved in the third stage. Efimov and the group [60] presented the Russian dataset for the QA system SberQuAD (Sberbank Question-Answering Dataset) in line with the SQuAD. In the context of the scarcity of non-English datasets, in general, the Russian dataset is specific for QA systems, and the SberQuAD can play a crucial role. Shelmanov and the group [61] applied semantic and semantic-syntactic methods to the Russian QA system. Both methods were applied separately and jointly, and later performance was compared, and the result showed that performance improved when the methods were applied jointly. Mochalova and Mochalov designed a multi-agent Russian QA system [62] where some agents (human beings) communicated in natural languages and some agents (computers and robots) used machine instructions. Such systems are beneficial for robotic communication. Later, they used semantic graphs to represent questions and sentences in the corpus. The concept of the semantic graph was extended with ontology, and onto semantic graph was used to retrieve the best answer to a question. The authors only proposed the algorithm in the paper [63]; they neither tested it to get the result nor compared it with other established methods.

5.6 LATEST RESEARCH WORKS ON QUESTION-ANSWERING SYSTEM IN MAJOR INDIAN LANGUAGES

As per articles 344(1) and 355, the eighth schedule of the constitution of India has 22 scheduled languages that have been given the status of official languages and are used by a considerably large population. Assamese, Bangla, Gujarati, Hindi, Kannada, Kashmiri, Konkani, Malayalam, Manipuri, Marathi, Nepali, Oriya, Punjabi, Sanskrit, Sindhi, Tamil, Telugu, Urdu, Bodo, Santhali, Maithili, and Dogri. This section explores recent research to develop QA systems in these scheduled languages. Bangla QA systems are discussed in the next section.

In Hindi, [64–76] is the current research on the QA system. Nanda and her group [64] used feature extraction and Naïve Bayes classification to determine the question category. The knowledge base is used to retrieve the sentence containing the answer. Devi and Dua [65] applied the KNN method for question classification and nine similarity functions to retrieve the answer. Smith-Waterman similarity gave the best result and handled both misspelled and multi-phase words. Ray and his group [66] presented an explorative review question-answering system in Hindi in their published book in 2018. In his master's thesis [67], Srivastava developed an architecture for the Hindi QA system using a PurposeNet-based ontology as the data resource. Bagde and her group [68] compared the performance of different similarity functions like "Jaro–Wrinkler," "Euclidean similarity measure," "Jaccard coefficient similarity," "N-gram approach", and "text similarity" in the case of the Hindi QA system. Schubotz and the group [69] developed a system to answer mathematical questions expressed in Hindi or English. Questions in natural language were processed and converted to the mathematical formula using the knowledge base of Wikidata. The values of the constants were also extracted from Wikidata. Singla and the group [70] compared the performance of different similarity functions in the bilingual QA system in Hindi and English. Bhagat, Prajapati, and Seth [71] developed an IVR-based automatic Hindi QA system in the domain named SRHR (sexual and reproductive health and rights). Taking the Jaccard similarity as a baseline, the authors applied the BERT model to the stored questions and answers to generate a QA pair and return it as the response to frequently asked questions (FAQ). Bhattacharyya and his group [72] developed a multi-domain, multilingual QA system in Hindi and English. Hindi questions are translated into English, and then CNN and RNN are applied to classify the questions. Candidate answers were extracted from the retrieved passages, which were retrieved by Lucene's text retrieval functionality and ranked using Boolean and BM25 vector space models. Finally, the prospective answers are ranked using different similarity functions. A good amount of accuracy is achieved in question classification. Gupta and his group [73] proposed a deep neural network-based model for a generalized multilingual QA system and tested the same for English-Hindi. The model has layers like question sentence encoding, probable answer sentence encoding, snippet encoding, and answer extraction. 39.44% exact match and 44.97 F1 scores are observed for the standard collected dataset, whereas 50.11% exact match and 53.77 F1 scores are observed in the translated SQuAD dataset. Chandu et al. [74] crowdsourced questions and answers for multilingual code-mixed languages. This has 1,694 Hinglish (Hindi+English) question-answer pairs. Though directly unrelated, Chandra and Dwivedi [75] presented an excellent way of query expansion using Okapi BM25 document ranking for Hindi queries in English document retrieval. Lewis et al. [76] presented a multilingual QA system for seven languages, including only Hindi as an Indian language. They presented the dataset and benchmark results by applying different ML and neural networking models to evaluate performance in multilingual forms in those seven languages.

The current QA system research is in the Malayalam language [77,78]. Archana and her group [77] developed a QA system in the Malayalam language using a supervised or rule-based approach. The approach includes Vibhakthi and POS tag analysis. Seena and her group [78] developed a Malayalam QA system that gives word-level answers by question type detection. After getting the type of question, the system finds prospective documents based on keyword matching using the "TnT tagger." Finally, the answer word is extracted using speculation and prediction from the question word. Overall, 70% accuracy is achieved by the system.

In the Marathi language, [79] and [80] are the current QA system research. Lende and Raghuwanshi [79] developed a QA system for the closed domain in the education sector. Though the system is claimed to be a generalized QA system, it is tested only in the Marathi QA system. Kamble and Baskar [80] manually translated the questions and answers of the famous television show "Kaun Banega Crorepati" or "KBC" into Marathi. Then, Marathi questions are classified using SVM. 73.5% accuracy is achieved in the best case.

In the Punjabi language, [81–83] is the current research on the QA system. Walia and the group [81] applied RNN and LSTM-based neural networks for scoring and ranking answers and observed improvements in results using statistical methods like cosine similarity, Jaccard similarity, etc. Dhanjal and the group [82] explored a gravity-based approach to building a QA system in the Punjabi language and observed 91% accuracy and 91% precision. Agarwal and Kumar [83] used Universal Networking

Language (UNL) to develop a QA system in Punjabi. The system achieved a 97.5 F1 score after UNLization of questions and answers in Punjabi.

Terdalkar and Bhattacharya [84] developed a QA system using a knowledge graph in Sanskrit. The system gives factoid answers on relationships based on the Ramayana, the Mahabharata, and the BhāvaprakāśaNighaṇṭu (an Ayurvedic text) with 50% accuracy.

In the Tamil language, [85–92] are the current research on QA systems. Ravi and Artstein [85] explored translation mechanisms from English to Tamil for dialogue systems. Performance is compared, and it is observed that the accuracy of correct response in Tamil is 79% and in English is 89%. Machine translation from English to Tamil drops accuracy by 54%, whereas Tamil to English stands at 79%. Niveditha [86] proposed the "Agrisage" QA system in Tamil based on HMM and ontology in her postgraduate thesis. The system is tested against 100 Tamil questions, and 68.33% precision is achieved. Sankaravelayuthan and the group [87] designed a parser for the Tamil QA system. The Stanford parser's POS tagging modules, chunking techniques, and dependency sets are modified to make them suitable for the Tamil QA system. Liu and the group [88] proposed a cross-lingual and multilingual QA system involving Tamil using a BERT-based neural networking model. Thara et al. [89] explored code-mixed QA systems using RNN and HAN (Hierarchical Attention Network) in Hindi, Telugu, and Tamil languages. On average, 80% accuracy is achieved in the best case. The work of Chandu et al. [90] is already mentioned for the Hinglish (Hindi+English) language. The same system for code-mixed languages works for the Tamlish (Tamil+English) language. Selvarasa et al. [91] explored various techniques for similarity calculation of short Tamil sentences, which is useful for ranking the answers. Knowledge-based methods and corpus-specific techniques blended with string similarity and graph alignment measurements gave 85% accuracy. Rajendran and the group [92] presented a beautiful resource: Ontological Structure of Tamil (OST). Based on Tamil WordNet and Tamil dictionary, this UGC-sponsored post-doctoral work will help Tamil's IR and QA systems.

In the Telugu language, [93–98] is the current research on the QA system. Ravva and her group [93] presented AVADHAN – the QA system in Telugu using SVM, multi-layer perceptrons, and logical regression. In the case of a partial match, SVM gave the highest accuracy of 68.5%. Khanam and Subbareddy [94] proposed a system for Telugu question-answering with structured databases using NLIDB (Natural Language Interface to Database). A query or question in Telugu natural language is transformed into a SQL query, and the answer is retrieved. Duggenpudi [95] built a QA system in Telugu using deep learning for question classification and a rule-based approach for processing the question. They got 99.326% accuracy in question classification using LSTM and 90.769% accuracy in retrieving the correct answer. Chandu et al. [96,97] are previously mentioned for Hindi and Tamil; the work includes a factoid QA system for the Telugu code-mixed (Telugu+English) language. Danda and the group [98] developed an end-to-end dialogue and QA system in Telugu in the tourism domain. They have claimed excellent accuracy for the system.

In Urdu, Singh et al. [99] proposed a cross-lingual QA system in 14 languages, including Urdu. BERT with a trained classification layer is used. Outstanding accuracy is achieved in four languages, including Urdu.

In Bangla, [100–115] are the current research on the QA system. Banerjee and his group [100–107] have primarily worked on the survey, dataset building in Bangla, and question classification. For question classification [102,103], with bagging and boosting of three primary classifiers, namely Naïve Bayes, Decision Tree, and Rule Induction, the system achieved 91.65% accuracy. [104] and [105] are proposed as the first resource for the Bangla QA system. Reference [106] is the presentation of work toward a code-mixed, cross script corpus. In [107], the system achieved 87.79% accuracy in question classification by combining four baseline classifiers. Monisha and the group [108,109] proposed question classification using SVM, Naïve Bayes, Stochastic Gradient Descent, and Decision Tree and retrieving answers using statistical parameters. Classification accuracy was 90.6%, and correct answer retrieval accuracy was 66.2% and 56.8%, respectively, in coarse and fine-grain levels. [110] is a good review from Anika et al. Bhuiyan et al. [111] proposed the Bangla QA system based on the attention mechanism. Kowsher and the group [112] classified the Bangla questions with a Naïve Bayes classifier and extracted the answer with cosine similarity and Jaccard similarity with 93.22% accuracy. Keya et al. [113] proposed a QA system with an LSTM-based deep learning method. Islam and the group [114] proposed SGD-based question classification and keyword-based answer retrieval. Rahman et al. [115] proposed CNN-based question classification with TF-IDF as the feature vector. In the next section, comparisons of the algorithm and results are presented in detail.

5.7 BACKEND DATABASE OR REPOSITORY USED IN THE QUESTION-ANSWERING SYSTEM

Digitization was first started in English. So, a large pool of English text was readily available. In addition, most of the websites were primarily in English. So, creating a repository through crawling or other curated methods was comparatively easy in English. The availability of text in English eased the job of researchers in getting a suitable repository for the QA system. In recent years, Stanford University created SQuAD [11–13], or Stanford Question-Answering Dataset, with the help of crowd workers. The volunteers contributed to building 100,000 questions and their answers in English on different topics from Wikipedia. This extremely useful resource from the Stanford NLP group is helping the scientists working on the English QA system by letting them concentrate solely on writing algorithms and building prediction models. English being the primitive language for communication on the internet, many software systems have been developed that have resulted in large structured or relational databases in English. In such cases, questions are transformed into SQL queries to give the correct answers. Nevertheless, semi-structured databases like DBPedia are also present in English, where questions are converted to SPARQL [7,9] queries to get the correct result. Some other languages, such as French [19], Portuguese [24], and Arabic [52], have also witnessed research in QA systems with a semi-structured database as a repository in recent years.

The scenario is not the same for Bangla or other Indian languages. Even 20 years ago, there was not enough web content in Bangla or other Indian languages. The situation has improved with the effort to present web content in regional languages. Still, there is a lack of large relational databases or semi-structured databases to experiment with the Bangla QA system. Initially, researchers invested their time in creating a curated Bangla corpus from various social media [100–107]. The same trend was found by creating a curated corpus from various Bangla websites [108–115] of the universities in Bangladesh. As the corpus is curated, the number of different domains or classes in the corpus is reduced. In the case of Bangla, all the repositories are found as flat text or doc files. Semi-structured or structured databases are unavailable in Bangla or are not large enough to carry out QA system research. Government of India has taken the initiative to build a standard corpus in different Indian languages under the project "Technology Development of Indian Languages" (TDIL) run by the Ministry of Electronics and IT, Govt. of India. The Bangla corpus is a collection of Bangla texts by eminent authors in 86 different domains. Still, the creation of questions and evaluation of the correctness of the retrieved answers are done with human effort in Bangla QA system research.

5.8 DIFFERENT APPROACHES OF ALGORITHMS USED IN BANGLA QA SYSTEM RESEARCH

Technically, any QA system is composed mainly of two phases. The first is question classification, and the second is answer retrieval. Question classification is required to determine the domain or category of the question in order to reduce the search space in the multi-domain repository. Answer retrieval includes searching for a prospective answer in the selected domain, and in many cases, it finds a list of answers based on some matching algorithms and ranks them according to their similarity score. Generally, one or more classification algorithms (often a combination of classifiers or an ensemble approach) are used to determine the category or domain of the question. A major thrust has been given to question classification implementation in the Bangla QA system. Some frameworks have been proposed for measuring similarity to get a list of prospective answers. Semantic similarity measurement and answer ranking are still in the initial phase in the case of the Bangla QA system. In Table 5.2, a synopsis of algorithms used in Bangla QA system research is presented.

Table 5.2 Existing algorithms in Bangla QA system

Citation	Proposed functionality	Algorithm	Dataset
[100, 101]	Bangla Question Classification	Naïve Bayes + Decision Tree	Curated corpus and questions (dataset) from BCSTAT. COM
[102, 103]	Bangla Question Classification	Ensemble Bagging and Boosting (Naïve Bayes + Kernel Naïve Bayes + Rule Induction + Decision Tree)	Same dataset as of [100,101]
[104, 105]	Resource Creation for Bangla QA	Bangla Factoid Question-Answering (BFQA) System architecture proposal and Bangla repository creation from Wikipedia	Own created dataset from Wikipedia
[106]	Resource Creation for Code-mixed Bangla QA	Corpus collection from various social media posts to build repository for code-mixed Bangla QA system	Own created dataset from social media posts
[107]	Bangla Question Classification	Combination of classifiers (Naïve Bayes + Kernel Naïve Bayes + Rule Induction + Decision Tree)	Same dataset as of [100,101]
[108]	Bangla Question Classification	Stochastic Gradient Descent + Decision Tree + Support Vector Machine + Naïve Bayes	Own created dataset from SUST website
[109]	Bangla Question Classification + Document identification for answer retrieval	Question Classification using the algorithm in [108] and CNN for document categorization	Own created dataset from SUST website

(Continued)

Table 5.2 (Continued) Existing algorithms in Bangla QA system

Citation	Proposed functionality	Algorithm	Dataset
[110]	Comparison of Question Classification algorithms	Comprehensive Comparison	NA
[111]	Bangla QA system	Bi-LSTM-based encoder and Attention Mechanism-based decoder	General knowledge-based question answers were collected from various Bangla websites
[112]	Establishment of a relationship between users' questions and prospective answers	Cosine similarity + Jaccard similarity + Naïve Bayes	Own created dataset from the NSTU website
[113]	Establishment of a relationship between users' questions and prospective answers	LSTM-based deep learning model	Curated General Knowledge dataset in Bangla
[114]	Question Classification and answer category detection	Stochastic Gradient Descent (SGD)–based classifier	Collected from the web. No explicit information is available
[115]	Bangla Question Classification	CNN for coarse and SGD for finer classification	Same dataset as of [114]

5.9 DIFFERENT ALGORITHMS USED IN BANGLA QA SYSTEM RESEARCH

For classification, different classifiers are used. Generally, classifiers are combined to get better results. Naïve Bayes, Kernel Naïve Bayes, Rule Induction, Decision Tree, Stochastic Gradient Descent, and SVM are the most frequently used classifiers, where the first four classifiers are combined by Banerjee et al. at Jadavpur University, Kolkata, India, with ensemble. The researchers used the last two classifiers in Shahjalal University of Science and Technology (SUST), Bangladesh, and Noakhali Science and Technology University (NSTU), Bangladesh. The training set is prepared from the corpus or repository sentences. The objective of the question classification is to reduce the search space and find the category of the question. After predicting the category, searching for the answer was done. To refine the processing further, in a few cases [115], two-stage classifications of the question are done. Two-stage classifications helped to reduce the search space further. The predicted document also needed to be reclassified in such cases to get the micro-level classes or taxonomies.

Research in answer retrieval is very limited in Bangla. Unlike classification, there is no common culture of research built up to retrieve and rank the answers. Locating the texts for the question and measuring the similarity between the question and the prospective answer is the initial step in answering. Then, predicted answers are ranked, or the actual answer is extracted from the only predicted text. Again, to reiterate, in the early stages of such research, the diversity of the nature and form of the corpus influenced the researchers to explore different options for getting an accurate answer. An architecture was proposed in BFQA [104,105] to retrieve the answer. A Bi-LSTM-based encoder and an attention mechanism-based decoder are used in [111].

5.10 RESULTS ACHIEVED BY VARIOUS BANGLA QA SYSTEMS

In this section, results from various Bangla QA systems are in Table 5.3. The objectives are also presented. As the dataset is different for different researchers, direct comparison is impossible. Still, it gives an idea of the performance required to achieve the desired goal for a different dataset with different algorithms.

Table 5.3 Comparison of results achieved by existing Bangla QA systems

Citation	Objective	Result	Comments
[100, 101]	Bangla question classification.	The accuracy of the classifiers is as follows. Naïve Bayes=81.89%, Kernel Naïve Bayes=83.21%, Rule Induction=85.57%, and Decision Tree=85.69%.	Raw classifiers were used in this early research in Bangla question classification without any ensemble or combination approach.
[102]	Bangla question classification with an ensemble approach.	Naïve Bayes=82.87%, Kernel Naïve Bayes=82.97%, Rule Induction=84.12%, and Decision Tree=88.21%.	The bagging and Boosting ensemble approach is used. The result is that Bagging and Decision Tree outperformed in Boosting as well.
[103]	Bangla question classification with different combination methods.	The best performance is achieved by the voting approach with 91.65% accuracy.	Combinations like an ensemble, stacking, and voting of classifiers are used.

(Continued)

Table 5.3 (Continued) Comparison of results achieved by existing Bangla QA systems

Citation	Objective	Result	Comments
[104]	BFQA: Bangla Factoid Question-Answering.	Mean Reciprocal Ranks for the answers are 34% for Geography and 31% for the Agriculture domain, and 32% overall.	The developed corpus is of two categories-Geography and Agriculture.
[105]	Resource creation for the Bangla QA system.	Not Applicable.	A corpus was created from original history, geography, and agriculture books to carry out future research in the Bangla QA system.
[106]	Resource creation for Benglish (Bangla+English) code-mixed QA system.	Not Applicable.	A corpus was created from the social media post to carry out future research in the Benglish code-mixed QA system.
[107]	Bangla question classification with different combination methods.	The stacking approach achieves the best performance with 87.79% accuracy for fine-grain classification.	Combinations like an ensemble, stacking, and voting of classifiers are used.
[108]	Bangla Question Classification.	90.6% accuracy in the best case.	Stochastic Gradient Descent+Decision Tree+Support Vector Machine+Naïve Bayes Classifiers are used, and the best result was achieved by SVM with linear kernel.
[109]	Bangla QA system with a statistical approach.	Accuracy of answer sentence (having the answer) extraction = 66.2% Accuracy of document (having the answer) extraction=72% Accuracy of question classification=90.6%	After question classification, a statistical approach was taken to retrieve the right answer.
[110]	Comparison of different classification-based machine learning techniques for Bangla question classification.	Multi-layer perceptron gave the best result with 83% accuracy.	This paper presents a comparison method between seven classifiers to classify the Bangla questions.

(Continued)

Table 5.3 (Continued) Comparison of results achieved by existing Bangla QA systems

Citation	Objective	Result	Comments
[111]	Bangla QA system.	The result is not reported.	Bi-LSTM-based encoder and Attention Mechanism-based decoder are used to train and test Bangla QA systems.
[112]	Establishment of a relationship between users' questions and prospective answers.	The percentage of accuracy for the correct retrieved answer is as follows for three methods- Cosine similarity=93.22%, Jaccard similarity=84.64%, and Naïve Bayes=91.31%.	Documents of 74 topics related to NSTU were used as corpus.
[113]	Bangla QA system using deep learning techniques.	89% accuracy is achieved.	LSTM-based deep learning model using seq2seq learning in the general knowledge dataset is used.
[114]	Question Classification and answer category detection using SGD-based classifier.	95.56% accuracy for coarse-grain classification and 87.64% accuracy for finer grain classification.	Syntactic features are used only.
[115]	Two-stage Bangla question classification.	95% accuracy for coarse-grain classification and 89% accuracy for finer grain classification.	CNN is used for coarse-grain classification, and SGD is used for finer grain classification.

5.11 CONCLUSION

This article presents the evolution of search and landmark events in the history of search engines. Next, different dimensions of "semantic analysis" research fields are discussed. Then, the catalyst events for research progress in the contextual research domains, specifically automatic QA, are described. Eventually, automatic QA proved to be the most promising and discrete field that evolved from search engines.

The industries are running to give semantically accurate responses to a question or query presented online. The later sections present recent and state-of-the-art research projects in the question-answering system in relevant languages.

Historically, QA systems have grown in English and later in other languages. This thesis concerns the improvement of semantic search in Indian languages, so the latest research in QA systems in various foreign languages related to Indian ethnicity is discussed. Then, the latest research on the QA system in major Indian languages is discussed. Lastly, all the valuable research in the Bangla QA system is presented. Repositories and algorithms used in different Bangla QA systems are discussed next.

The survey found that the research work in the Bangla QA system is still not mature. There is some memorable research in question classification in the case of the Bangla QA system, but there is almost nil or very little research on semantic answer retrieval and ranking for a Bangla question. That is the significant research gap in the case of the Bangla QA system.

REFERENCES

[1] Green Jr, B.F., Wolf, A.K., Chomsky, C. and Laughery, K., 1961, May. Baseball: an automatic question-answerer. In *Papers presented at the May 9–11, 1961, Western Joint IRE-AIEE-ACM Computer Conference* (pp. 219–224). Association for Computing Machinery, Los Angeles, California.

[2] Woods, W.A., 1973, June. Progress in natural language understanding: an application to lunar geology. In *Proceedings of the June 4–8, 1973, National Computer Conference and Exposition* (pp. 441–450). Association for Computing Machinery, New York.

[3] https://en.wikipedia.org/wiki/ELIZA (as on 07.08.2020 at 8 PM).

[4] https://www.ibm.com/ibm/history/ibm100/us/en/icons/watson/ (as on 07.08.2020 at 10 PM).

[5] Gao, T., Fodor, P. and Kifer, M., 2019. Querying knowledge via multi-hop english questions. *Theory and Practice of Logic Programming*, 19(5–6), pp. 636–653.

[6] Liu, Y., Yi, X., Chen, R., Zhai, Z. and Gu, J., 2018. Feature extraction based on information gain and sequential pattern for English question classification. *IET Software*, 12(6), pp. 520–526.

[7] Radoev, N., Zouaq, A., Tremblay, M. and Gagnon, M., 2018, June. A language adaptive method for question answering on French and English. In *Semantic Web Evaluation Challenge* (pp. 98–113). Springer, Cham.

[8] Moore, K., 2019. *Building an Automated QA System Using Online Forums as Knowledge Bases* (Electronic Theses and Dissertations, University of Mississippi, 1692). https://egrove.olemiss.edu/etd/1692.

[9] Diefenbach, D., Both, A., Singh, K. and Maret, P., 2020. Towards a question answering system over the semantic web. Semantic Web, (Preprint), pp. 1–19.

[10] Liu, Y., Zhang, C., Yan, X., Chang, Y., and Yu, P.S., 2019, November. Generative question refinement with deep reinforcement learning in retrieval-based QA system. In *Proceedings of the 28th ACM International Conference on Information and Knowledge Management* (pp. 1643–1652). Association for Computing Machinery, Beijing, China.

[11] Rajpurkar, P., Zhang, J., Lopyrev, K. and Liang, P., 2016. Squad: 100,000+ questions for machine comprehension of text. arXiv preprint arXiv:1606.05250.

[12] Rajpurkar, P., Jia, R. and Liang, P., 2018. Know what you don't know: unanswerable questions for SQuAD. arXiv preprint arXiv:1806.03822.

[13] https://rajpurkar.github.io/SQuAD-explorer/ (as on 15.09.2020 at 7 PM).

[14] Vaswani, A., Shazeer, N., Parmar, N., Uszkoreit, J., Jones, L., Gomez, A.N., Kaiser, Ł. and Polosukhin, I., 2017. Attention is all you need. In *Advances in Neural Information Processing Systems* (pp. 5998–6008). Neural Information Processing Systems Foundation, Inc. (NeurIPS), California.

[15] Lan, Z., Chen, M., Goodman, S., Gimpel, K., Sharma, P. and Soricut, R., 2019. Albert: A lite bert for self-supervised learning of language representations. arXiv preprint arXiv:1909.11942.

[16] Zhang, Z., Yang, J. and Zhao, H., 2020. Retrospective reader for machine reading comprehension. arXiv preprint arXiv:2001.09694.

[17] Segura, C., Palau, À., Luque, J., Costa-Jussà, M.R. and Banchs, R.E., 2019. Chatbol, a chatbot for the Spanish "La Liga". In *9th International Workshop on Spoken Dialogue System Technology* (pp. 319–330). Springer, Singapore.

[18] Carrino, C.P., Costa-jussà, M.R. and Fonollosa, J.A., 2019. Automatic Spanish translation of the squad dataset for multilingual question answering. arXiv preprint arXiv:1912.05200.

[19] Radoev, N., Tremblay, M., Gagnon, M. and Zouaq, A., 2017, May. AMAL: answering French natural language questions using DBpedia. In *Semantic Web Evaluation Challenge* (pp. 90–105). Springer, Cham.

[20] Keraron, R., Lancrenon, G., Bras, M., Allary, F., Moyse, G., Scialom, T., Soriano-Morales, E.P. and Staiano, J., 2020. Project PIAF: building a native French question-answering dataset. arXiv preprint arXiv:2007.00968.

[21] Falke, S., 2019, June. Developing a knowledge graph for a question answering system to answer natural language questions on German grammar. In *European Semantic Web Conference* (pp. 199–208). Springer, Cham.

[22] da Conceição Rodrigues, R.M., 2017. RAPPORT: a fact-based question answering system for Portuguese (Doctoral dissertation, Universidade de Coimbra, Portugal).

[23] Rodrigues, R. and Gomes, P., 2016, July. Improving question-answering for Portuguese using triples extracted from corpora. In *International Conference on Computational Processing of the Portuguese Language* (pp. 25–37). Springer, Cham.

[24] de Araujo, D.A., Rigo, S.J., Quaresma, P. and Muniz, J.H., 2020, March. A Portuguese dataset for evaluation of semantic question answering. In *International Conference on Computational Processing of the Portuguese Language* (pp. 217–227). Springer, Cham.

[25] Damiano, E., Spinelli, R., Esposito, M. and De Pietro, G., 2016, January. Towards a framework for closed-domain question-answering in Italian. In *2016 12th International Conference on Signal-Image Technology & Internet-Based Systems (SITIS)* (pp. 604–611). IEEE, Naples, Italy.

[26] Croce, D., Zelenanska, A. and Basili, R., 2018, November. Neural learning for question answering in Italian. In *International Conference of the Italian Association for Artificial Intelligence* (pp. 389–402). Springer, Cham.

[27] Polignano, M., Basile, P., de Gemmis, M., Semeraro, G. and Basile, V., 2019, November. AlBERTo: Italian BERT language understanding model for NLP challenging tasks based on tweets. In *CLiC-it* (pp. 1–6). CEUR Workshop Proceedings (CEUR-WS.org), Bari, Italy.

[28] Cuteri, B., Reale, K. and Ricca, F., 2019, May. A logic-based question answering system for cultural heritage. In *European Conference on Logics in Artificial Intelligence* (pp. 526–541). Springer, Cham.

[29] Gunawan, A.A., Mulyono, P.R. and Budiharto, W., 2018. Indonesian question answering system for solving arithmetic word problems on intelligent humanoid robot. *Procedia Computer Science, 135*, pp. 719–726.

[30] Putra, S.J., Naf'an, M.Z. and Gunawan, M.N., 2018, July. Improving the scoring process of question answering system in Indonesian language using fuzzy logic. In *2018 International Conference on Information and Communication Technology for the Muslim World (ICT4M)* (pp. 239–242). IEEE, Kuala Lumpur, Malaysia.

[31] Day, M.Y., Tsai, C.C., Chuang, W.C., Lin, J.K., Chang, H.Y., Sun, T.J., Tsai, Y.J., Chiang, Y.H., Han, C.Z., Chen, W.M. and Tsai, Y.D., 2016, June. IMTKU question answering system for world history exams at NTCIR-12 QA Lab2 task. In *NTCIR* (pp. 425–431). National Institute of Informatics, Tokyo, Japan.

[32] Sakamoto, K., Ishioroshi, M., Matsui, H., Jin, T., Wada, F., Nakayama, S., Shibuki, H., Mori, T. and Kando, N., 2016. Forst: question answering system for second-stage examinations at NTCIR-12 QA Lab-2 task. In *NTCIR* (pp. 467–472). National Institute of Informatics, Tokyo, Japan.

[33] Kato, S., Togashi, R., Maeda, H., Fujita, S. and Sakai, T., 2017, August. LSTM vs. BM25 for open-domain QA: a hands-on comparison of effectiveness and efficiency. In *Proceedings of the 40th International ACM SIGIR Conference on Research and Development in Information Retrieval* (pp. 1309–1312). National Institute of Informatics, Tokyo, Japan.

[34] Liu, X., Chen, Q., Deng, C., Zeng, H., Chen, J., Li, D. and Tang, B., 2018, August. LCQMC: a large-scale Chinese question matching corpus. In *Proceedings of the 27th International Conference on Computational Linguistics* (pp. 1952–1962). Association for Computational Linguistics, Santa Fe, New Mexico.

[35] Chen, Z., Yin, S. and Zhu, X., 2020, April. Research and implementation of QA system based on the knowledge graph of Chinese classic poetry. In *2020 IEEE 5th International Conference on Cloud Computing and Big Data Analytics (ICCCBDA)* (pp. 495–499). IEEE, Chengdu, China.

[36] Zhou, X., Wu, B. and Zhou, Q., 2018. A depth evidence score fusion algorithm for Chinese medical intelligence question answering system. *Journal of Healthcare Engineering, 2018*, 1205354.

[37] Tian, B., Zhang, Y., Chen, X., Xing, C. and Li, C., 2019, April. DRGAN: a GAN-based framework for doctor recommendation in Chinese on-line QA communities. In *International Conference on Database Systems for Advanced Applications* (pp. 444–447). Springer, Cham.

[38] Liu, J., Yang, Y., Lv, S., Wang, J. and Chen, H., 2019. Attention-based BiGRU-CNN for Chinese question classification. *Journal of Ambient Intelligence and Humanized Computing*, 1–12. https://doi.org/10.1007/s12652-019-01344-9

[39] Tsai, H.W., 2018. Extending Knowledge in Different Languages with the Deep Learning Model for a Chinese QA System.

[40] Guo, H., Na, X. and Li, J., 2018. Qcorp: an annotated classification corpus of Chinese health questions. *BMC Medical Informatics and Decision Making*, *18*(1), p. 16.

[41] Feng, G., Du, Z. and Wu, X., 2018. A Chinese question answering system in medical domain. *Journal of Shanghai Jiaotong University (Science)*, *23*(5), pp. 678–683.

[42] Liang, H., Lin, K. and Zhu, S., 2019, October. Short text similarity hybrid algorithm for a Chinese medical intelligent question answering system. In *National Conference on Computer Science Technology and Education* (pp. 129–142). Springer, Singapore.

[43] Liu, H.I., Ni, C.C., Hsu, C.H., Chen, W.L., Chen, W.M. and Liu, Y.T., 2020, February. Attention based R&CNN medical question answering system in Chinese. In *2020 International Conference on Artificial Intelligence in Information and Communication (ICAIIC)* (pp. 341–345). IEEE.

[44] Zhang, Y., Lu, W., Ou, W. et al., 2020. Chinese medical question answer selection via hybrid models based on CNN and GRU. *Multimed Tools Appl* 79, pp. 14751–14776. https://doi.org/10.1007/s11042-019-7240-1.

[45] Sun, Y., Chen, C., Xia, T., and Zhao, X., 2019. QuGAN: quasi generative adversarial network for Tibetan question answering corpus generation. *IEEE Access*, *7*, pp. 116247–116255.

[46] Sun, Y. and Xia, T., 2019. A hybrid network model for Tibetan question answering. *IEEE Access*, *7*, pp. 52769–52777.

[47] Zhu, S., He, X., La, B. and Yu, H., 2019, June. Research on Tibetan question answering technology based on deep learning. *Journal of Physics: Conference Series*, *1213*(2), p. 22006.

[48] Mozannar, H., Hajal, K.E., Maamary, E. and Hajj, H., 2019. Neural Arabic question answering. arXiv preprint arXiv:1906.05394.

[49] Hasan, A.M., Rassem, T.H. and Noorhuzaimi, M.N., 2018, June. Combined support vector machine and pattern matching for Arabic Islamic hadith question classification system. In *International Conference of Reliable Information and Communication Technology* (pp. 278–290). Springer, Cham.

[50] Dardour, S., Fehri, H. and Haddar, K., 2019, June. Disambiguation for Arabic question answering system. In *International Conference on Automatic Processing of Natural-Language Electronic Texts with NooJ* (pp. 101–111). Springer, Cham.

[51] Hasan, A.M., Rassem, T.H. and Karimah, M.N., 2018. Pattern-matching based for Arabic question answering: a challenge perspective. *Advanced Science Letters*, *24*(10), pp. 7655–7661.

[52] Bouziane, A., Bouchiha, D., Doumi, N. and Malki, M., 2018. Toward an Arabic question answering system over linked data. *Jordanian Journal of Computers and Information Technology (JJCIT)*, *4*(2), pp. 102–115.

[53] Al-Shenak, M.O.A.Y.E.A.H., Nahar, K. and Halwani, H., 2019. AQAS: Arabic question answering system based on SVM, SVD, and LSI. *Journal of Theoretical and Applied Information Technology*, *97*(2), pp. 681–691.

[54] Ismail, W.S. and Homsi, M.N., 2018. Dawqas: a dataset for Arabic why question answering system. *Procedia Computer Science*, *142*, pp. 123–131.

[55] Razzaghnoori, M., Sajedi, H. and Jazani, I.K., 2018. Question classification in Persian using word vectors and frequencies. *Cognitive Systems Research*, *47*, pp. 16–27.

[56] Boreshban, Y., Yousefinasab, H. and Mirroshandel, S.A., 2018. Providing a religious corpus of question answering system in Persian. *Signal and Data Processing*, *15*(1), pp. 87–102.

[57] Boreshban, Y. and Mirroshandel, S.A., 2017. A novel question answering system for religious domain in Persian. *Electronics Industries*, 8(2), pp. 73–88.

[58] Nikolaev, K. and Malafeev, A., 2017, July. Russian-language question classification: a new typology and first results. In *International Conference on Analysis of Images, Social Networks and Texts* (pp. 72–81). Springer, Cham.

[59] Soboleva, D. and Vorontsov, K., 2019. Three-stage question answering system with sentence ranking. *EPiC Series in Language and Linguistics*, *4*, pp. 18–25.

[60] Efimov, P., Chertok, A., Boytsov, L. and Braslavski, P., 2020, September. Sberquad-Russian reading comprehension dataset: description and analysis. In *International Conference of the Cross-Language Evaluation Forum for European Languages* (pp. 3–15). Springer, Cham.

[61] Shelmanov, A.O., Kamenskaya, M.A., Ananyeva, M.I. and Smirnov, I.V., 2017. Semantic-syntactic analysis for question answering and definition extraction. *Scientific and Technical Information Processing*, *44*(6), pp. 412–423.

[62] Mochalova, A. and Mochalov, V., 2017, October. Multi-agent question-answering system. In *International Conference on BioGeoSciences* (pp. 29–39). Springer, Cham.

[63] Mochalova, A. and Mochalov, V., 2018, August. Evaluation of the similarity coefficient of the ontosemantic graphs and its application for the problem of question-answering search. In *2018 3rd Russian-Pacific Conference on Computer Technology and Applications (RPC)* (pp. 1–6). IEEE, Vladivostok, Russia.

[64] Nanda, G., Dua, M. and Singla, K., 2016, March. A Hindi question answering system using machine learning approach. In *2016 International Conference on Computational Techniques in Information and Communication Technologies (ICCTICT)* (pp. 311–314). IEEE, New Delhi, India.

[65] Devi, R. and Dua, M., 2016. Performance evaluation of different similarity functions and classification methods using web-based Hindi language question answering system. *Procedia Computer Science*, *92*, pp. 520–525.

[66] Ray, S.K., Ahmad, A. and Shaalan, K., 2018. A review of the state of the art in Hindi question answering systems. In *Intelligent Natural Language Processing: Trends and Applications* (pp. 265–292). Springer, Cham.

[67] Srivastava, R., 2017. PurposeNet Ontology-based Question Answering (QA) System for Hindi (Doctoral dissertation, International Institute of Information Technology Hyderabad).

[68] Sneha, B., Mohit, D. and Singh, V.Z., 2016. Comparison of different similarity functions on Hindi QA system. In *Proceedings of International Conference on ICT for Sustainable Development* (pp. 657–663). Springer, Singapore.

[69] Schubotz, M., Scharpf, P., Dudhat, K., Nagar, Y., Hamborg, F. and Gipp, B., 2018. *Introducing MathQA: A Math-Aware Question Answering System.* Information Discovery and Delivery.

[70] Singla, K., Dua, M. and Nanda, G., 2016, March. A language-based comparison of different similarity functions and classifiers using web-based bilingual question answering system developed using machine learning approach. In *Proceedings of the Second International Conference on Information and Communication Technology for Competitive Strategies* (pp. 1–4). Association for Computing Machinery, Udaipur, India.

[71] Bhagat, P., Prajapati, S.K. and Seth, A., 2020, June. Initial lessons from building an IVR-based automated question-answering system. In *Proceedings of the 2020 International Conference on Information and Communication Technologies and Development* (pp. 1–5). Association for Computing Machinery, Guayaquil, Ecuador.

[72] Gupta, D., Kumari, S., Ekbal, A. and Bhattacharyya, P., 2018, May. MMQA: a multi-domain multilingual question-answering framework for English and Hindi. In *Proceedings of the Eleventh International Conference on Language Resources and Evaluation (LREC 2018)* (pp. 2777–2784). European Language Resources Association (ELRA), Miyazaki, Japan.

[73] Gupta, D., Ekbal, A. and Bhattacharyya, P., 2019. A deep neural network framework for english hindi question answering. *ACM Transactions on Asian and Low-Resource Language Information Processing (TALLIP)*, 19(2), pp. 1–22.

[74] Chandu, K., Loginova, E., Gupta, V., Genabith, J.V., Neumann, G., Chinnakotla, M., Nyberg, E. and Black, A.W., 2019. Code-mixed question answering challenge: Crowd-sourcing data and techniques. In *Third Workshop on Computational Approaches to Linguistic Code-Switching* (pp. 29–38). Association for Computational Linguistics, Melbourne, Australia.

[75] Chandra, G. and Dwivedi, S.K., 2019. Query expansion for effective retrieval results of Hindi-English cross-lingual IR. *Applied Artificial Intelligence*, 33(7), pp. 567–593.

[76] Lewis, P., Oğuz, B., Rinott, R., Riedel, S. and Schwenk, H., 2019. MLQA: Evaluating cross-lingual extractive question answering. arXiv preprint arXiv:1910.07475.

[77] Archana, S.M., Vahab, N., Thankappan, R. and Raseek, C., 2016. A rule-based question answering system in Malayalam corpus using vibhakthi and pos tag analysis. *Procedia Technology*, 24, pp. 1534–1541.

[78] Seena, I.T., Sini, G.M. and Binu, R., 2016. Malayalam question answering system. *Procedia Technology*, 24, pp. 1388–1392.

[79] Lende, S.P. and Raghuwanshi, M.M., 2016, February. Question answering system on education acts using NLP techniques. In *2016 World Conference on Futuristic Trends in Research and Innovation for Social Welfare (Startup Conclave)* (pp. 1–6). IEEE, Coimbatore, India.

[80] Kamble, S. and Baskar, S., 2018. Learning to classify Marathi questions and identify answer type using machine learning technique. In *Advances in Machine Learning and Data Science* (pp. 33–41). Springer, Singapore.

[81] Walia, T.S., Josan, G.S. and Singh, A., 2019. An efficient automated answer scoring system for Punjabi language. *Egyptian Informatics Journal, 20*(2), pp. 89–96.

[82] Dhanjal, G.S., Sharma, S. and Sarao, P.K., 2016. Gravity based Punjabi question answering system. *International Journal of Computer Applications, 147*(3), p. 21.

[83] Agarwal, V. and Kumar, P., 2018. UNLization of Punjabi text for natural language processing applications. *Sādhanā, 43*(6), p. 87.

[84] Terdalkar, H. and Bhattacharya, A., 2019. Framework for question-answering in Sanskrit through automated construction of knowledge graphs. In *Proceedings of the 6th International Sanskrit Computational Linguistics Symposium* (pp. 97–116). Association for Computational Linguistics, Kharagpur, India.

[85] Ravi, S. and Artstein, R., 2016, September. Language portability for dialogue systems: translating a question-answering system from English into Tamil. In *Proceedings of the 17th Annual Meeting of the Special Interest Group on Discourse and Dialogue* (pp. 111–116). Association for Computational Linguistics, Los Angeles.

[86] Karmegam, N., 2019. Agrisage: an HMM and ontology-based cross-lingual question answering system for the agricultural domain (Doctoral dissertation).

[87] Sankaravelayuthan, R., Anandkumar, M., Dhanalakshmi, V. and Mohan Raj, S.N., 2019. A Parser for Question-answer System for Tamil. *QA System Using DL, 229*, p. 230.

[88] Liu, J., Lin, Y., Liu, Z. and Sun, M., 2019, July. XQA: a cross-lingual open-domain question answering dataset. In *Proceedings of the 57th Annual Meeting of the Association for Computational Linguistics* (pp. 2358–2368). Association for Computational Linguistics, Florence, Italy.

[89] Thara, S., Sampath, E. and Reddy, P., 2020, June. Code mixed question answering challenge using deep learning methods. In *2020 5th International Conference on Communication and Electronics Systems (ICCES)* (pp. 1331–1337). IEEE, Coimbatore, India.

[90] Chandu, K., Loginova, E., Gupta, V., Genabith, J.V., Neumann, G., Chinnakotla, M., Nyberg, E. and Black, A.W., 2019. Code-mixed question answering challenge: Crowd-sourcing data and techniques. In Third Workshop on Computational Approaches to Linguistic Code-Switching (pp. 29–38). Association for Computational Linguistics, Melbourne, Australia.

[91] Selvarasa, A., Thirunavukkarasu, N., Rajendran, N., Yogalingam, C., Ranathunga, S. and Dias, G., 2017, May. Short Tamil sentence similarity calculation using knowledge-based and corpus-based similarity measures. In *2017 Moratuwa Engineering Research Conference (MERCon)* (pp. 443–448). IEEE, Moratuwa, Sri Lanka.

[92] Rajendran, S., Soman, K.P., Anandkumar, M. and Sankaralingam, C., 2020. Ontological structure-based retrieval system for Tamil. In *Applications in Ubiquitous Computing* (pp. 197–223). Springer, Cham.

[93] Ravva, P., Urlana, A. and Shrivastava, M., 2020. AVADHAN: system for open-domain Telugu question answering. In *Proceedings of the 7th ACM IKDD CoDS and 25th COMAD* (pp. 234–238).

[94] Khanam, M.H. and Subbareddy, S.V., 2017. Question answering system with natural language interface to database. *International Journal of Management, IT and Engineering*, 7(4), pp. 38–48.

[95] Duggenpudi, S.R., Varma, K.S.S. and Mamidi, R., 2019, November. Samvaadhana: a Telugu dialogue system in hospital domain. In *Proceedings of the 2nd Workshop on Deep Learning Approaches for Low-Resource NLP (DeepLo 2019)* (pp. 234–242). Association for Computational Linguistics, Hong Kong, China.

[96] Chandu, K.R., Chinnakotla, M., Black, A.W. and Shrivastava, M., 2017, September. Webshodh: a code mixed factoid question answering system for web. In *International Conference of the Cross-Language Evaluation Forum for European Languages* (pp. 104–111). Springer, Cham.

[97] Chandu, K., Loginova, E., Gupta, V., Genabith, J.V., Neumann, G., Chinnakotla, M., Nyberg, E. and Black, A.W., 2019. Code-mixed question answering challenge: crowd-sourcing data and techniques. In *Third Workshop on Computational Approaches to Linguistic Code-Switching* (pp. 29–38). Association for Computational Linguistics, Melbourne, Australia.

[98] Danda, P., Jwalapuram, P. and Shrivastava, M., 2017, December. End to end dialog system for Telugu. In *Proceedings of the 14th International Conference on Natural Language Processing (ICON-2017)* (pp. 265–272). NLP Association of India, Kolkata, India.

[99] Singh, J., McCann, B., Keskar, N.S., Xiong, C. and Socher, R., 2019. Xlda: Cross-lingual data augmentation for natural language inference and question answering. arXiv preprint arXiv:1905.11471.

[100] Banerjee, S. and Bandyopadhyay, S., 2012, December. Bengali question classification: Towards developing QA system. In *Proceedings of the 3rd Workshop on South and Southeast Asian Natural Language Processing* (pp. 25–40). The COLING 2012, Mumbai, India.

[101] Bandyopadhyay, SBS, 2012. Bengali question classification: towards developing qa system. In *24th International Conference on Computational Linguistics* (p. 25–40). The COLING 2012 Organizing Committee, Mumbai, India.

[102] Banerjee, S. and Bandyopadhyay, S., 2013. Ensemble approach for fine-grained question classification in Bengali. In *27th Pacific Asia Conference on Language, Information, and Computation* (pp. 75–84). The COLING 2012 Organizing Committee, Mumbai, India.

[103] Banerjee, S. and Bandyopadhyay, S., 2013, October. An empirical study of combing multiple models in Bengali question classification. In *Proceedings of the Sixth International Joint Conference on Natural Language Processing* (pp. 892–896). Asian Federation of Natural Language Processing, Nagoya, Japan.

[104] Banerjee, S., Naskar, S.K. and Bandyopadhyay, S., 2014, September. BFQA: a Bengali factoid question answering system. In *International Conference on Text, Speech, and Dialogue* (pp. 217–224). Springer, Cham.

[105] Banerjee, S., Lohar, P., Naskar, S.K. and Bandyopadhyay, S., 2014, September. The first resource for Bengali question answering research. In *International Conference on Natural Language Processing* (pp. 290–297). Springer, Cham.

[106] Banerjee, S., Naskar, S.K., Rosso, P. and Bandyopadhyay, S., 2016, March. The first cross-script code-mixed question answering corpus. In *MultiLingMine@ ECIR* (pp. 56–65). CEUR Workshop Proceedings, Padova, Italy

[107] Banerjee, S., Naskar, S.K., Rosso, P. and Bandyopadhyay, S., 2019. Classifier combination approach for question classification for Bengali question answering system. *Sādhanā*, *44*(12), p. 247.

[108] Monisha, S.T.A., Sarker, S. and Nahid, M.M.H., 2019, May. Classification of Bengali questions towards a factoid question answering system. In *2019 1st International Conference on Advances in Science, Engineering and Robotics Technology (ICASERT)* (pp. 1–5). IEEE, Dhaka, Bangladesh.

[109] Sarker, S., Monisha, S.T.A. and Nahid, M.M.H., 2019, September. Bengali question answering system for factoid questions: a statistical approach. In *2019 International Conference on Bangla Speech and Language Processing (ICBSLP)* (pp. 1–5). IEEE, Sylhet, Bangladesh.

[110] Anika, A., Rahman, M., Islam, D., Jameel, A.S.M.M. and Rahman, C.R., 2019. Comparison of machine learning-based methods used in Bengali question classification. *arXiv preprint arXiv:1911.03059*.

[111] Bhuiyan, M.R., Masum, A.K.M., Abdullahil-Oaphy, M., Hossain, S.A. and Abujar, S., 2020, July. An approach for Bengali automatic question answering system using attention mechanism. In *2020 11th International Conference on Computing, Communication and Networking Technologies (ICCCNT)* (pp. 1–5). IEEE, Kharagpur, India.

[112] Kowsher, M., Rahman, M.M., Ahmed, S.S. and Prottasha, N.J., 2019, December. Bangla intelligence question answering system based on mathematics and statistics. In *2019 22nd International Conference on Computer and Information Technology (ICCIT)* (pp. 1–6). IEEE, Dhaka, Bangladesh.

[113] Keya, M., Masum, A.K.M., Majumdar, B., Hossain, S.A. and Abujar, S., 2020, July. Bengali question answering system using Seq2Seq learning based on general knowledge dataset. In *2020 11th International Conference on Computing, Communication and Networking Technologies (ICCCNT)* (pp. 1–6). IEEE, Kharagpur, India.

[114] Islam, M.A., Kabir, M.F., Abdullah-Al-Mamun, K. and Huda, M.N., 2016, May. Word/ phrase-based answer type classification for Bengali question answering system. In *2016 5th International Conference on Informatics, Electronics and Vision (ICIEV)* (pp. 445–448). IEEE, Dhaka, Bangladesh.

[115] Rahman, M.H., Rahman, C.R., Amin, R., Sifat, M.H.R. and Anika, A., 2020, February. A hybrid approach towards two-stage Bengali question classification utilizing smart data balancing technique. In *International Conference on Cyber Security and Computer Science* (pp. 454–464). Springer, Cham.

Chapter 6

Recent advances in textual code-switching

Sergei Ternovykh
Rostelecom Contact Center

Anastasia Nikiforova
Smartcat

6.1 INTRODUCTION

Code-switching is a linguistic phenomenon of shifting from one language or language variant to another and back within one spoken or written utterance. Today, over 7,000 languages are spoken across the world. However, what is more, speakers often mix languages in everyday communication, especially given that more than half of the world's population is fluent in two or more languages. As research shows, code-switching is present in almost any form of conversation involving multilingualism, whether it is a mere use of borrowed words and phrases or the intentional use of several languages. Such shifts allow the speakers to convey messages more precisely, emphasize specific ideas, establish group identity, and reduce social and interpersonal distance.

Code-switching is peculiar to multilingual regions where speakers are fluent in two or more languages - Texas and California in the US, Hong Kong, and Macau in China, and most countries in Europe, Africa, and Southeast Asia. Also, multi-lingualism is very common in India, where there are around 500 spoken languages, with about 30 languages with over 1 million speakers. Despite geography, code-switching is also triggered by speakers' backgrounds and the topic of the conversation.

Code-switching is typically considered a feature of informal communication, such as casual speech or online texting. However, there is a plethora of evidence that language alteration also happens in formal situations, for instance, in newspaper headlines, political speeches, and teaching. Thus, code-switching is a natural form of communication and should be broadly analyzed.

The tendency to research code-switching as an NLP problem has emerged with the increased popularity of online communication and the abundance of textual data. The texts crawled online are often unstructured and contain mixed-language texts. Typically, a straightforward approach to word-level language identification (LID) gives unsatisfactory results because the

DOI: 10.1201/9781003244332-6

speakers tend to write using only one script, usually Roman, for simplicity. Below is an example of how the speaker used Roman script instead of Japanese syllabary in a Japanese-English code-switched sentence.

1. Kore wa *cat* ya.
   (There is a *cat*.)

We first introduce the code-switching phenomenon from a theoretical linguistic perspective and describe the typology of code-switching most often used in NLP research. Next, we reflect on why code-switching is an essential and relevant NLP problem. The following section lists the most popular datasets with code-switching to solve various NLP tasks. Afterward, we describe recent technologies used to build code-switching models in various NLP applications. Then, we explain why the evaluation of code-switching is a challenging task and describe recently introduced evaluation benchmarks. Finally, we contemplate current research limitations and possible future research directions.

6.2 BACKGROUND

The term 'code-switching' is often referred to as 'code-mixing'. Although some linguists insist there is a difference between code-switching and code-mixing, in our survey, we use these terms interchangeably, as in most NLP papers.

6.2.1 Theoretical approaches and types of code-switching

There are several linguistic approaches to code-switching. One of them, a constraint-based model by Poplack [1], suggests that code-switching is subject to two constraints. The first, *free-morpheme constraint*, implies that language alterations cannot occur between a lexical stem and bound morphemes, distinguishing code-switching from borrowing. The second, *equivalence constraint*, stipulates that shifts only happen at the point where the surface structures of the languages concur. In other words, the code-switched utterance must obey the grammatical rules of both languages, as, for example, in the following English-Spanish sentence.

2. I like you *porque eres amable.*
   (I like you *because you are kind.*)

Depending on where the shift happens in the utterance, Poplack distinguishes three types of code-switching - *tag-switching*, *intrasentential switching*, and *intersentential switching*.

Tag-switching implies the insertion of idiomatic phrases, exclamations, or fillers from one language into an utterance in another language. For example, in the following Indonesian sentence, the speaker uses an English exclamation:

3. *Oh, God!* Saya lupa nggak bawa ban serep.
   (*Oh God!* I forgot I did not bring a spare tire.)

Intrasentential switching happens when a phrase in one language is used within a sentence in another language. In the following sentence, the speaker switches from French to Wolof.

4. *Le matin de bonne heure* ngay joge Medina pour dem juilli.
   (*Early in the morning*, you leave Medina to go to pray.)

Finally, intersentential switching occurs between sentences, with each being monolingual, as shown in the following Turkish-English sentence.

5. Onun için çok böyle birkaç ay çok rahatsiz oldum okulda. *There was almost no communication.*
   (That is why it was so uncomfortable for a few months at school. *There was almost no communication.*)

Poplack's constraint-based model initially considered only close languages such as English and Spanish. The model was criticized as insufficiently restrictive and not applicable to many languages and cases of code-switching. For instance, the possible intra-word switching contradicts the free-morpheme constraint, and switching postpositional phrases in Hindi with prepositional English phrases contradicts the equivalence constraint.

6. But ma-*day*-s a-no a-ya ha-ndi-si ku-mu-on-a.
   (But these *days*, I do not see him much.) Shona-English intra-word switching.
7. John gave a book *ek larakii ko.*
   (John gave a book *to a girl.*) Prepositional English phrase in place of a postpositional Hindi phrase.

Despite the criticism, Poplack's classification into tag-switching, intrasentential switching, and intersentential switching is often used in NLP papers, along with another approach – Carol Myers-Scotton's Matrix Language Frame (MLF) model [2,3].

According to the MLF, code-switching implies one primary language (*matrix language, ML*) and a secondary language (or *embedded language, EL*). In written communication, the EL may be either inserted into the ML phrase (intrasentential switch) or the two codes may be used separately, alternating after clauses or sentences (intersentential switch). In the case of an intrasentential switch, the distribution of the two languages is asymmetrical, with ML being dominant. Supposedly, ML is the language that is more common for the speaker.

There are two principles to identify the ML. According to the *Morpheme-Order Principle*, if any number of ML morphemes surround a single EL lexeme, the surface order will be that of ML. The *System Morpheme Principle*, in turn, states that all the system morphemes (i.e., morphemes that build the grammatical frame of the word) will come from the ML.

Within the MLF model, Myers-Scotton also developed the Markedness model, one of the most full-fledged sociolinguistic theories of the motivation behind code-switching. It states that speakers *choose* to code-switch to index the rights and obligations relative to their interlocutors in the conversation set. If the set of rights and obligations is expected and unmarked, the choice is marked if it is different. Speakers make marked choices, usually accompanied by prosodic features such as pauses or metacommentary, to consciously establish their social position, whereas unmarked choices are used to explore possible language choices.

According to [4], the ML in the utterance can be approximated based on the overall word count and verb and functional word frequencies. The assumption is that whichever language has the most significant word, verb, and functional word counts combined with the ML.

6.2.2 Measuring code-switching complexity

Various metrics have been proposed to measure the amount of code-switching in the corpora.

6.2.2.1 The code-mixing index

Code-mixing index (CMI) [5] measures the amount of switching in the utterance or the whole corpus based on word frequency.

$$
CMI = \frac{\sum_{i=1}^{N} (w_i) - max\{w_i\}}{n - u},
$$

where $\sum_{i=1}^{N} (w_i)$ is the sum over all N languages of their respective number of words, $max\{w_i\}$ is the highest word count from any language, n is the total number of tokens, and u is the total number of language-independent (i.e., they are the same for all languages involved) tokens.

At the utterance level, it amounts to finding the most frequent language in the utterance and counting the frequency of the words belonging to all other involved languages. The metric is calculated as an average of all non-zero CMIs of each utterance at the corpus level.

6.2.2.2 The multilingual index

Multilingual index (M-index) [6] calculates the ratio of languages based on the Gini coefficient to measure the unevenness of language distribution in the corpus.

$$M\text{-}index = \frac{1 - \sum_{j=1}^{N} p_j^2}{(N-1) \cdot \sum_{j=1}^{N} p_j^2},$$

where $N > 1$ is the number of languages represented in the corpus and p_j is the total count of all the tokens in the language j divided by the number of words in the corpus. The final M-index ranges between 0 (for monolingual corpora) and 1 (all languages are distributed equally).

6.2.2.3 CMI and M-index

CMI and M-index are useful for determining the language ratio in code-switching corpora, but they do not show the frequency of switches. The Integration Index (I-Index) [7] measures how often languages alternate.

$$I\text{-}Index = \frac{1}{n-1} \sum_{1 \le i = j-1 \le n-1}^{n} S(l_i, l_j),$$

It is calculated as a proportion of switch point counts to the number of language-dependent tokens in the corpus; hence, it shows the probability of any given token being a switch point, where the ratio $1/(n-1)$ reflects that there are overall $(n-1)$ possible switch points in a corpus of size n, and $S(l_i, l_j) = 1$ if $l_i \neq l_j$ and 0 otherwise. The I-Index ranges from 0 for monolingual texts to 1 for texts where any two consecutive language-dependent tokens are from different languages.

6.2.2.4 Burstiness

Burstiness, proposed by [8], goes beyond simple word count by analyzing the temporal distribution of code-switching across the corpus and comparing it with the Poisson process. Applying this metric to code-switching allows us to see whether languages alternate sporadically (as in the Poisson process) or periodically. Given σ_τ (the standard deviation of the language spans) and m_τ (the mean of the language spans), the burstiness metric is measured as follows

$$Burstiness = \frac{(\sigma_\tau/m_\tau - 1)}{(\sigma_\tau/m_\tau + 1)} = \frac{(\sigma_\tau - m_\tau)}{(\sigma_\tau + m_\tau)}$$

and it lies within the [−1,1] range, where 1 is texts where code-switching happens at random and −1 are texts with frequent code-switching, i.e., switching patterns are more predictable.

6.3 CODE-SWITCHING DATASETS

In recent years, with the rising popularity of deep neural networks for solving practically any NLP task, there has been an explosion of available labeled datasets. Consequently, several code-switching datasets are available for download and training custom models for solving problems such as part-of-speech (POS) tagging, Named Entity Recognition, LID, and many more. Sure enough, usually, those are datasets for the most high-resource languages, including English, Mandarin, Hindi, and Modern Arabic. Languages with lower textual resources available for crawling are still underrepresented; however, every year, there are more.

Here we list and observe textual datasets for various language pairs concerning the NLP problem they aim to solve.

6.3.1 Language identification

The corpora listed below are available for training LID models on code-switched data. The labels in these datasets are at the token level.

- In [9], the authors propose a large mixed-language dataset where each text contains English and one or several of 30 other languages. As separating languages written in different scripts would have been trivial, the authors only considered texts where all the words were written in Latin characters. The corpus is designed for training LID models.
- The first workshop on computational approaches to code-switching released corpora for Mandarin-English, Modern Standard Arabic-Dialectal Arabic, Nepali-English, and Spanish-English language pairs [10].
- Four corpora were released for The First Shared Task on Language Identification in Code-Switched Data, including Modern Standard Arabic and Dialectal Arabic, Mandarin-English, Nepali-English, and Spanish-English language pairs [11]. The texts were crawled from Twitter, Facebook, and various forums and blogs.
- Another two corpora were released for The Second Shared Task on Language Identification in Code-Switched Data from Modern Standard Arabic and Dialectal Arabic and Spanish-English language pairs [12]. All the data were collected from Twitter: 9,979 tweets for Modern Standard Arabic and Dialectal Arabic, and 10,716 for Spanish-English.

- Reference [13] introduced a Malayalam-English code-mixed dataset collected from YouTube comments with word-level language annotation.

6.3.2 POS tagging

Below is the list of corpora with annotated POS tags on code-switched data. Most of these corpora are also annotated for LID.

- Solorio and Liu released 922 sentences (~8,000 words) of transcribed English-Spanish spoken conversations with manually annotated POS tags and 239 detected language alterations [14]. The authors combined tagsets from the English and Spanish Tree Taggers for corpus annotation.
- In reference [15], the authors present a code-mixed English-Hindi corpus gathered from Twitter (4,435 tweets) and Facebook (1,236 posts). The corpus is aimed at POS tagging: 1,106 texts from the corpus – 552 Facebook posts and 554 tweets – were manually annotated with POS tags, language ids, and utterance boundaries. For tokenization, the authors decided to use the CMU tokenizer, initially designed for English but showing decent results for Hindi. The CMI for Twitter+Facebook utterances is 21.68 for non-zero CMI utterances (68% of the corpora).
- A corpus of English-Hindi tweets is more significant than previously available corpora presented in reference [16]. The corpus consists of 98,000 tweets, of which 1500 random samples are manually annotated with language ids and a universal POS tagset [17]. All the tweets collected are related to five specific events.
- A corpus of Hindi-English code-mixed social media text is presented in [18]. The authors explore LID, back-transliteration, normalization, and POS tagging of the code-switched data. The corpus comprises 6,983 posts and comments (1,13,578 tokens), semi-automatically cleaned and formatted. Moreover, the authors added three kinds of named entity tags – people, location, and organization, because they occurred in almost every text.
- In the course of the 12th International Conference on Natural Language Processing [19], organizers released datasets with code-switched texts from three language pairs: Bengali-English (4,296 sentences), Hindi-English (1,106 sentences), and Tamil-English (918 sentences). Reference [20] proposes a POS-annotated version of these datasets.
- A Turkish-German dataset with Universal Dependencies POS tagset annotation and language ids is presented in [21]. The corpus includes 1,029 tweets, with at least one code-switching point in each of them.

- The Miami Bangor Corpus [22] includes 56 transcribed conversations in Spanish-English from speakers living in Miami, FL. The original POS labeling was automatically annotated and had many drawbacks; therefore, [23] proposes POS tagging the corpus based on the crowd-sourced tagset.

6.3.3 Named entity recognition

Like LID and POS tagging, code-switched corpora aimed at NER have word-level annotations.

- A corpus of Hindi-English code-switched tweets on topics such as politics, social events, sports, and others is introduced in [24]. The named entity tagset consists of four categories: Person, Organization, Location, and Other, and the annotation is based on BIO-standard.
- Two corpora were presented during the third shared task of the Computational Approaches to Linguistic Code-Switching (CALCS) workshop [25]: English-Spanish (50,727 tweets) and Modern Standard Arabic-Egyptian (10,102 tweets). All the texts were collected from Twitter, and tokens were annotated with nine types of named entities.

6.3.4 Chunking and dependency parsing

The following code-switching corpora are labeled with chunking tags or syntactic dependency information.

- In reference [26], the authors performed dependency parsing on Komi-Zyrian and Russian code-switching texts. The authors emphasize that their approach was explicitly concerned with low-resource languages, to which Komi-Zyrian belongs. Thus, they trained the dependency model on monolingual data from both languages and applied the cross-lingual results to code-switched data. The training part of the corpus is 3,850 Russian texts from Universal Dependencies v2.0, and the Komi part is only 40 manually collected sentences. The test set consists of 80 Komi-Russian code-switched sentences.
- Duong et al. annotated the NLmaps corpus [27], a semantic parsing corpus for English and German, using transfer learning: the model was trained on monolingual English and German data with no cases of code-switching and used to annotate German-English code-switched sentences [28].
- A dataset of 450 Hindi-English code-mixed tweets from Hindi multilingual speakers is presented [29]. The dataset is manually annotated with Universal Dependencies and language ids.
- In [30], the authors annotated 8,450 Hindi-English media texts with a coarser version of the AnnCorra chunking tagset [31]. Chunking, or shallow parsing, was modeled as three different sequence labeling

problems: chunk label, chunk boundary, and combined label. The additional labels include universal POS tagset, language ids, and normalized script.

6.3.5 Sentiment analysis

Another category of code-mixed datasets includes sentiment annotations at the sentence or utterance level.

- The first code-switching English-Spanish corpus with sentiment labels is introduced in [32]. The dataset comprises 3,062 tweets, and each is manually annotated according to the SentiStrength strategy [33], i.e., each tweet was assigned a dual score (p,n) from 1 to 5, where p and n indicate the positive and negative sentiments. The authors provide a second labeling strategy in [34], converting their original annotation to the trinary polarity categories: positive, neutral, and negative.
- A Hindi-English code-mixed corpus for sentiment analysis is described in [35]. The texts are posts and comments collected from two popular Facebook pages – one of a famous Indian actor and another of a current Indian Prime Minister. Overall, there are 3,859 code-mixed texts in the corpus.
- The authors of [36] collected an English-Chinese code-mixed dataset from a Chinese microblogging website *Weibo.com*. The tagset consists of five basic emotions: *happiness, sadness, fear, anger,* and *surprise*. Out of 4195 annotated posts, 2,312 posts express one or more of these emotions.

Apart from that, there are also corpora with hate speech and irony annotations, similar in meaning to sentiment labels.

- A Twitter corpus of Hindi-English code-mixed tweets with Hate Speech annotation is introduced in [37]. Out of 4,575 tweets, hate speech was found in 1,661 of them.
- In [38], the authors release a Hindi-English code-mixed social media corpus with irony annotation. The dataset of 3,055 code-switched tweets is annotated with language ids at the token level and *Ironic/ Non-Ironic* labels at the tweet level.

6.3.6 Question-answering

Code-switched question-answering (QA) datasets typically consist of code-mixed questions and answers.

- Chandu et al. manually prepared a dataset with 1,000 code-mixed Hindi-English questions based on an Indian adaptation of the popular "Who Wants to Be a Millionaire?" TV show [39]. The authors emphasize that even though the third of all tokens is in English, the syntactic structure of the questions is characteristic of Hindi.

- 5,933 questions for Hindi-English, Tamil-English, and Telugu-English based on images and articles are collected in [40]. The final dataset described included 1,694 Hindi-English, 2,848 Tamil-English, and 1,391 Telugu-English factoid questions and their answers.
- The authors of [41] used a monolingual English SimpleQuestions dataset and weakly supervised bilingual embeddings, resulting in 250 Hindi-English code-switched questions.
- Around 300 messages from social media are leveraged in [42] to collect 506 questions around sports and tourism domains.

6.3.7 Conversational systems

One of the goals of studying code-switching as an NLP problem is to build a conversational system that can understand code-mixed languages. The datasets below were collected with this goal in mind.

- The authors of [43] were the first to compile code-switched goal-oriented conversations for four Indian languages alternating with English: Hindi, Bengali, Gujarati, and Tamil. With the help of crowd-sourcing, they used the DSTC2 restaurant reservation dataset and a mix of code-switched corpora to build a sizable 5-way parallel code-switched corpora.
- Following this, [44] constructed two Hindi-English corpora for goal-oriented conversational systems. Content writers manually compiled the first dataset and included 7,000 code-mixed samples transformed from the English Snips dataset. The second dataset combines the manually created code-switched set and two parallel monolingual parts - original English utterances and their automatic Hindi translations provided by Google Translate. Overall, the second dataset includes 21,000 training samples.

6.3.8 Machine translation

Research shows that incorporating code-mixing knowledge helps improve machine translation (MT) quality [45–47]. The corpora below were designed to help MT models understand code-switched inputs.

- A new parallel corpus with 6,096 code-switched Hindi-English sentences on one side and their English translation on the other is collected in [48]. The dataset was used to augment the existing machine translation systems, improving translation quality.

- Another Hindi-English parallel corpus containing code-mixed posts and comments from Facebook is released in [48]. The dataset also considers the spelling variants in social media code-switching, e.g., '*plzzz*' instead of '*please.*'
- A parallel multilingual code-switching corpus from UN documents is collected [50]. The source texts are a code-mix of Arabic, mainly English, French, Spanish, and other languages, and the translation side is English.

6.3.9 Natural language inference

Natural language inference (NLI) is the NLP task of inferring a logical relationship between a premise and a hypothesis, such as entailment or contradiction. Reference [51] presents the first dataset for code-switched NLI, where both the premises and hypotheses are in code-switched Hindi-English. The premises are compiled from Bollywood Hindi movies, and hypotheses are crowdsourced from Hindi-English bilingual speakers. The corpus includes 400 code-switched conversation snippets as premises and 2,240 code-switched hypotheses.

6.4 NLP TECHNIQUES FOR TEXTUAL CODE-SWITCHING

The authors of [52], a brief survey of code-switching studies in NLP, provide an extensive survey of code-switching applications in NLP (Figure 6.1). Each NLP task requires a different approach, especially when training on code-switched data.

In reference [53], the authors outline the main challenges in the computational processing of core NLP tasks and downstream applications. They highlight problems at different stages - data collection, normalization, language modeling, LID, various annotations, and, finally, building a model that can deal with code-switching.

The recent tendency in feature extraction from texts is to use various word embeddings. If natural data is scarce, code-switched embeddings can be trained with either natural code-switched corpora or artificially synthesized code-switched data. However, the most recent approach uses large multilingual models such as multilingual BERT [54] and other transformer-based models. Recently, reference [55] suggested fine-tuning multilingual BERT on code-switched data as an effective way to improve BERT's code-switching understanding performance.

The recent tendency in feature extraction from texts is to use various word embeddings. If natural data is scarce, code-switched embeddings can be trained with either natural code-switched corpora or artificially

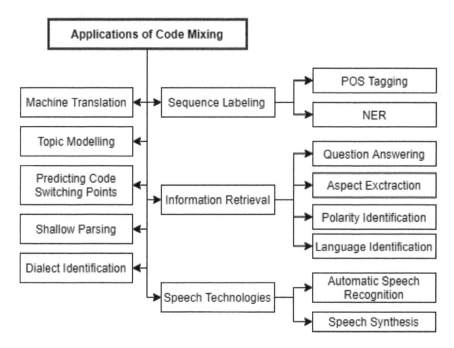

Figure 6.1 Applications of code-mixing. (Based on [52].)

synthesized code-switched data. However, the most recent approach uses large multilingual models such as multilingual BERT [54] and other transformer-based models. Recently, reference [55] suggested fine-tuning multilingual BERT on code-switched data as an effective way to improve BERT's code-switching understanding performance.

This section provides a comprehensive description of recent advances in textual code-switched NLP. Various approaches have been proposed to build NLP systems on code-switched data, depending on the languages, data availability, and task specifics.

6.4.1 Language modeling

Language models (LMs) are used in a variety of NLP systems. Although many textual code-switched datasets are described in Section 3, robust LMs usually require millions of sentences to train, which makes building LMs for code-switched languages challenging, especially for low-resource languages. When training data is scarce, some approaches use only monolingual data to train LMs, while others combine extensive monolingual data with a small set of code-mixed texts.

Some approaches implement grammatical constraints based on theories of code-switching to constrain search paths in LMs using artificially generated data. The authors of [56] use *equivalence constraints* to predict code-switching points. Language-independent *Functional Head Constraints* (FHC) are integrated with [57] for code-switching into the Mandarin-English LM. FHC theory states that no code-switch is allowed between a functional head (complementizer, determiner, inflection, etc.) and its complement (sentence, noun-phrase, verb-phrase) [58].

A pipeline for English-Spanish code-switching where a recurrent neural network (RNN) LM is trained on monolingual data in both languages followed by code-mixed texts is presented in [59]. Furthermore, this pipeline is extended by applying grammatical models of code-switching to generate artificial code-switched texts and using a small subset of natural code-switched texts to sample from the artificially generated data to build improved LMs [60].

Research shows that incorporating various grammatical information helps improve LMs. In [61], the authors show that encoding language information allows LMs to predict code-switch points. In [62], the authors suggest a discriminative training approach for modeling code-mixed data. Alternatively, [63] employs clustering for the infrequent words to create an n-gram-based LM. In [64], the authors jointly performed multitask learning by training a LM and a POS tagging model. Lee and Li train a bilingual attention LM that learns cross-lingual probabilities from parallel texts and the language modeling objective, significantly reducing model perplexity [65].

6.4.2 Language identification

LID is one of the essential tasks for lexical-level modeling in downstream NLP applications. Most approaches focus on token-level LID; however, a few works perform LID at the utterance level. For instance, reference [66] used an utterance-level LID system based on the two languages' language ratio.

In reference [67], the authors of the Bengali, Hindi, and English code-switching dataset explore three LID approaches: (1) heuristic dictionary lookup, (2) word-level classification without contextual information using supervised machine learning with SVMs, and (3) SVMs and sequence labeling using CRFs, both employing contextual information. The best result was achieved using the CRF model.

One of the first computational approaches toward determining intra-word code-switching through segmenting tokens into smaller but meaningful morphological units and using them in a CRF model is suggested in reference [68].

In reference [69], the authors use language usage patterns of Hindi-English and consecutive POS tag information for LID. An LID system named EQUILID was introduced in reference [70]. They modeled switching with character-based sequence-to-sequence models to support dialectal and multilingual language varieties.

Recently, POS tagging has also been examined as a means to perform LID in code-mixed texts [71]. The authors collected a Devanagari corpus and annotated it with POS labels. All the texts in Devanagari were then transliterated into Roman script. The complementary English data is annotated with POS tags as well. The authors experimented with classical machine learning approaches, including SVM, decision trees, logistic regression, and random forests. The feature set included POS labels, word lengths, and words. The highest performance was achieved with random forest on word, word length, and POS label features.

6.4.3 POS tagging

The first problem with POS tagging code-switched text is that the languages involved are likely to have different POS tagsets due to differences in grammar. Many studies use a combination of the two or universal tagsets, such as [17], to overcome this problem.

A two-step mechanism for POS tagging Hindi-English texts is used in [18]. The LID step was performed using a pre-trained word-level logistic regression classifier. Then, they used an open-source CRF++-based POS tagger for Hindi and the Twitter POS tagger for English.

A language-specific POS tagging is performed in [15] on the custom Hindi-English dataset using classic ML approaches, namely CRFs, Sequential Minimal Optimization, Naive Bayes, and Random Forests. The authors show that the best results were achieved with Random Forests.

In reference [72], the authors use a stacked model technique and compare it to joint modeling and pipeline-based approaches. They show that the best-stacked model that utilizes all features outperforms the joint- and pipeline-based models. Later, reference [73] normalizes code-switched texts and presents their positive impact on POS tagging results.

The authors of [74] investigate POS tagging in two code-switching language pairs: Spanish-English and Modern Standard Arabic-Arabic dialects. The authors compare the use of two POS taggers versus a unified SVM tagger trained on code-mixed data. The best results are achieved with the SVM model.

A POS tagging algorithm for the transliterated text that does not require a separate LID step is described in reference [75]. The BiLSTM model uses pre-trained Hindi-English word vectors (indic-word2vec). The authors claim their approach outperforms standard models trained on monolingual text without transliteration.

In reference [76], the authors build a deep learning model for Modern Standard Arabic-Egyptian Arabic augmented with linguistic features, including POS labels. The authors also show that in intrasentential code-switching, POS categories constrain the selection of lexical items: function words are most likely to come from the EL, while content words typically come from the ML.

6.4.4 Named entity recognition

When alternating from one language to another, it is common for speakers to highlight certain named entities using switches. This phenomenon was also addressed during the Third Workshop on Computational Approaches to Linguistic Code-Switching (CALCS 2018). In the shared task, [77] improved state-of-the-art character-level Convolutional Neural Networks (CNNs) with BiLSTMs followed by a CRF layer by enriching data from external sources, stacking layers of pre-trained embeddings, Brown clusters, and a named entity gazetteer. Reference [78] built a model using character-based BiLSTM architecture. Prior to CALCS 2018, [79] suggested using character-level CNNs to model non-standard spelling variations, followed by a word-level Bi-LSTM to model sequences. The authors also highlight the importance of using predefined named entity gazetteer lists.

A code-switched social media corpus to benchmark for the task of Named Entity Recognition in Arabic-English is collected in [80]. The corpus is composed of 6,525 sentences that contain 1,36,574 tokens. They train a BiLSTM-CRF model, an extensively studied architecture for solving NER. They experimented with various word embeddings from the Flair framework [81].

To mitigate the high percentage of out-of-vocabulary words in code-switched data, [82] extends the LSTM architecture with bilingual character representations using transfer learning. The authors also denoise the data with spelling normalization.

As an alternative to the fusion approach described above, [83] uses the self-attention mechanism over character embeddings and feeds the resulting output to a BiLSTM model with residual connections. Inspired by this approach, [84] proposes multilingual meta-embeddings (MME) as an effective method to learn multilingual representations by leveraging monolingual pre-trained embeddings. They bypass the lexical-level LID problem using the exact self-attention mechanism on pre-trained word embeddings. For English-Spanish NER, [85] proposes Hierarchical meta-embeddings (HME) that learn to combine multiple monolingual word-level and subword-level embeddings to create language-independent lexical representations.

6.4.5 Dependency parsing

A neural approach for predicting arc-eager transitions on Hindi-English data by leveraging pre-existing monolingual annotated resources for training is proposed in [29]. Consequently, [30] proposes a shallow parsing or chunking model for Hindi-English code-mixed data. They train a CRF model to separately predict chunk labels, chunk boundaries, and combination of the two. The feature set includes POS labels and a word lemma.

In [27], the authors claim they were the first to perform semantic parsing on code-mixed data. They used transfer learning on English-German data, utilizing cross-lingual word embeddings in a sequence-to-sequence framework.

A Universal Dependencies dataset in Hindi-English is presented in [86] with a feature-level neural stacking model for dependency parsing and a three-step decoding scheme that outperforms prior approaches.

6.4.6 Sentiment analysis

Based on a benchmarking English-Spanish code-switching dataset [34], the same group of authors then compares a multilingual model trained on a multilingual dataset, separate monolingual models, and a monolingual model with incorporated LID [32]. The research demonstrates the effectiveness of the multilingual model in code-mixed scenarios.

A more classical approach based on word probabilities is presented in [87] for Telugu-English code-mixed data. Specifically, the model predicts tweet sentiment by transliterating each word in Roman script to the corresponding Telugu script and measuring word probability in each class.

During the shared task for sentiment analysis in Hindi-English and Bengali-English social media texts [88], the best-performing system combined word- and character-level n-gram features fed into an SVM classifier. Similarly, [89] achieved the best results using SVM for sentiment analysis of movie reviews in Bengali-English.

More recent studies use deep learning instead of classic machine learning algorithms. Thus, [90] feeds CNN subword-level representations to a dual encoder, presumably capturing the sentiment at both the sentence and subword levels. Likewise, [91] builds a CNN-based encoder for stance detection in Hindi-English code-mixed tweets.

In [92], the authors address hate speech detection in Hindi-English code-switched tweets. Employing transfer learning, they first train a CNN-based model on a large dataset of hateful tweets as a source task, followed by fine-tuning on transliterated texts in the same language. In [93], the authors combine basic linguistic and psycholinguistic features and feed them to a deep learning ensemble containing Deep Pyramid CNN, Pooled BiLSTM, and Disconnected RNN with concatenated Glove and FastText embeddings. The final decision is then made based on model averaging. In reference [94], the authors investigate subworld-level LSTM and Hierarchical LSTM to detect hate speech from code-mixed data.

6.4.7 Natural language inference

NLI is a relatively new and challenging problem for code-switched NLP with insufficient available annotated data. The task is to predict if a hypothesis entails or contradicts the given logical premise. Khanuja et al. present the first work on code-switching NLI based on Bollywood movies' conversations [51]. The authors performed fine-tuning of a multilingual BERT for the task, but the resulting accuracy of this model was relatively low, showing that NLI on code-switched data requires future work and more extensive training data.

6.4.8 Machine translation

In [95], the authors show that a zero-shot Neural Machine Translation model can also deal with code-switched inputs. The authors do not change the standard attention-based encoder-decoder NMT architecture but instead introduce an artificial token at the beginning of the input text to specify the required target language. Using a shared word piece vocabulary enables code-mixed NMT using a single model without increasing model parameters.

6.4.9 Question-answering

Code-switching-sensitive QA is a very impactful downstream NLP application, especially in domains such as health and technology, where the vocabulary rapidly grows and changes, resulting in variations of usage and code-mixing scenarios.

One of the first attempts to leverage code-mixed data to perform QA was [39]. The authors used monolingual English questions from an Indian adaptation of the famous "Who Wants to Be a Millionaire?" TV show. They then used crowdsourcing to convert these questions into Hindi-English code-mixed variants. The resulting dataset is transformed into a feature vector and passed through a one-vs-rest SVM model, which outputs a class with the maximum score.

A two-phase approach to QA is adopted in [41]: candidate generation and a Triplet-Siamese-Hybrid CNN (TSHCNN) to re-rank candidate answers. The model is trained on English questions from the SimpleQuestions dataset and noisy Hindi translations of these questions. The resulting network can effectively answer code-mixed English-Hindi questions, eliminating the need for translation into English.

6.5 EVALUATION OF CODE-SWITCHED SYSTEMS

Most of the progress in code-switching NLP was shaped by shared tasks in which common datasets are released and participants compete to build state-of-the-art systems for solving a specific task. Over the last few years, there have been several shared tasks and workshops for code-switched text processing. The tasks included LID [10–12], code-mixed entity extraction and information retrieval [96–98], POS tagging [99], named entity recognition [25], sentiment analysis [88], and QA [40].

These shared tasks have stimulated research in their respective subfields of code-switched NLP. Nevertheless, until recently, it was not clear how well these individual models could generalize across different tasks and languages. Two evaluation benchmarks for code-switching across different NLPs were proposed to fill this gap.

The GLUECoS benchmark [100] includes LID from text, POS tagging, named entity recognition, sentiment analysis, QA, and a new task for code-switching, NLI. It comprises 11 datasets across Spanish-English and Hindi-English language pairs, and the authors express their readiness to improve the benchmark by adding more diverse tasks and language pairs in a future version.

The LinCE benchmark [101] consists of 10 datasets across five language pairs. The tasks include LID, named entity recognition, POS tagging, and sentiment analysis.

Evaluations on the GLUECoS and LinCE benchmarks show that multilingual contextual LMs such as multilingual BERT significantly outperform cross-lingual word embedding models. Moreover, these models can be further improved by adding artificially generated code-switched data to the pre-training stage, as described in [100].

These evaluation benchmarks also show that although models on word-level tasks such as LID and named entity recognition perform highly, their performance on downstream tasks such as QA and NLI is poor. It is also clear that the models show significantly lower results on code-switched data than on monolingual corpora. This indicates that large multilingual models such as BERT do not perform as well on code-mixed data as on monolingual or cross-lingual tasks, but fine-tuning these models on code-switched data appears to be a promising future research direction.

6.6 CURRENT LIMITATIONS AND FUTURE WORK

Currently, code-switching is seen as a bridge to anchor representations in different languages to bring them closer and improve performance in cross-lingual NLP tasks. As a result, it would become possible to build models that can understand and interact with bilingual speakers similarly.

The main limitation of code-switching NLP is the lack of annotated data. Code-switching texts are primarily characteristic of casual spoken language and are significantly harder to collect than monolingual corpora. Moreover, as most studies focus on pairs where one language is high-resource (e.g., English, Spanish, and Modern Standard Arabic) and the other is low-resource, research on code-mixing as an NLP problem will always be data starved and require unique approaches for working under limited resources.

To alleviate the data sparsity problem, there are various techniques to generate code-switched text artificially, for instance, a code-switching-aware

RNN decoder [102] or a machine translation system to create code-switched data from monolingual data [103]. However, such techniques are not popular as they can be time-consuming and require a deep understanding of code-switching patterns.

Given that code-switching is mostly about informal communication close to spoken language, we can assume that some NLP tasks and topics are more likely to involve code-mixing than others. Intuitively, legal documents or parliamentary transcripts are likely monolingual, whereas social media texts and TV show transcripts may include code-switching, especially in multilingual regions. Moreover, some issues arise when processing colloquial text - from non-canonicity and misspelling to out-of-vocabulary words to incomplete syntactic structures.

Currently, most models for code-switching text generation rely on the syntactic constraints of the language pair. However, it is also essential to research the possibility of including sociolinguistic features of code-switching into NLP systems. Building models incorporating these factors could lead to better natural language understanding and text generation, which could enhance the quality of chatbots and dialogue systems.

The lack of standardized datasets makes it challenging to evaluate code-switched NLP models. Despite the recently introduced GLUECoS and LinCE benchmarks, the field needs a comprehensive evaluation of code-switched systems across various NLP tasks in many language pairs. Extensive evaluation benchmarks are even more critical due to the rising popularity of zero-shot cross-lingual transfer learning models, which could show good results when low-resource languages are involved.

Code-switching NLP is not yet a mature and well-researched field. Despite substantial work in individual subfields of code-switching NLP, there are no end-to-end systems capable of understanding and interacting in code-switched language with multilingual speakers. Learning to deal with data scarcity and incorporating sociolinguistic factors would benefit the resulting code-switching NLP models.

REFERENCES

[1] Poplack, S. (1980). Sometimes I'll start a sentence in Spanish y termino en español: toward a typology of code-switching. *Linguistics* 18: 581–618.

[2] Myers-Scotton, C. (1993). *Duelling Languages: Grammatical Structure in Codeswitching*. Oxford: Oxford University Press.

[3] Myers-Scotton, C., & Jake, J.L. (1995). Matching lemmas in a bilingual language competence and production model: evidence from intrasentential code-switching. *Linguistics* 33: 981–1024.

[4] Bullock, B.E., Guzman, W., Serigos, J., Sharath, V., & Toribio, A.J. (2018). Predicting the presence of a Matrix Language in code-switching. CodeSwitch@ ACL.

[5] Gambäck, B., & Das, A. (2016). Comparing the level of code-switching in Corpora. LREC.

[6] Barnett, R., Codó, E., Eppler, E., Forcadell, M., Gardner-Chloros, P., Hout, R., Moyer, M.G., Torras, M., Turell, M.T., Sebba, M., Starren, M., & Wensing, S. (2000). The LIDES coding manual: a document for preparing and analyzing language interaction data version.

[7] Guzmán, G.A., Serigos, J., Bullock, B.E., & Toribio, A.J. (2016). Simple tools for exploring variation in code-switching for linguists. CodeSwitch@ EMNLP.

[8] Goh, K., & Barabasi, A. (2006). Burstiness and memory in complex systems. EPL, 81.

[9] King, B., & Abney, S.P. (2013). *Labeling the Languages of Words in Mixed-Language Documents using Weakly-Supervised Methods*. NAACL.

[10] Diab, M., Hirschberg, J., Fung, P., & Solorio, T., Eds. (2014). *Proceedings of the First Workshop on Computational Approaches to Code Switching*. Doha, Qatar: Association for Computational Linguistics.

[11] Solorio, T., Blair, E., Maharjan, S., Bethard, S., Diab, M., Ghoneim, M., Hawwari, A., AlGhamdi, F., Hirschberg, J., Chang, A., & Fung, P. (2014). Overview for the first shared task on language identification in codeswitched data. *Proceedings of the First Workshop on Computational Approaches to Code Switching*. Doha, Qatar: Association for Computational Linguistics, pp. 62–72.

[12] Molina, G., Alghamdi, F., Ghoneim, M.A., Hawwari, A., Rey-Villamizar, N., Diab, M.T., & Solorio, T. (2016). Overview for the second shared task on language identification in code-switched data. ArXiv, abs/1909.13016.

[13] Thara, S. & Poornachandran, P. (2021). *Corpus Creation and Transformer based Language Identification for Code-Mixed Indian Language*. IEEE Access, pp. 1–14.

[14] Solorio, T., & Liu, Y. (2008). *Learning to Predict Code-Switching Points*. EMNLP, pp. 973–981.

[15] Jamatia, A., Gambäck, B., & Das, A. (2015). *Part-of-Speech Tagging for Code-Mixed English-Hindi Twitter and Facebook Chat Messages*. RANLP, pp. 239–248.

[16] Singh, K., Sen, I., & Kumaraguru, P. (2018). A Twitter corpus for Hindi-English code mixed POS tagging. SocialNLP@ACL, pp. 12–17.

[17] Petrov, S., Das, D., & McDonald, R.T. (2012). A universal part-of-speech tagset. ArXiv, abs/1104.2086.

[18] Vyas, Y., Gella, S., Sharma, J., Bali, K., & Choudhury, M. (2014). POS tagging of English-Hindi code-mixed social media content. EMNLP, pp. 974–979.

[19] *Proceedings of the 12th International Conference on Natural Language Processing, ICON 2015*, Trivandrum, India, December 11–14, 2015.

[20] Ghosh, S., Ghosh, S., & Das, D. (2016). Part-of-speech tagging of code-mixed social media text. CodeSwitch@EMNLP, pp. 90–97.

[21] Çetinoglu, Ö., & Çöltekin, Ç. (2016). Part of speech annotation of a Turkish-German code-switching corpus. LAW@ACL.

[22] Deuchar, M., Webb-Davies, P., Herring, J., Parafita Couto, M.C., & Carter, D.M. (2014). Building bilingual corpora. *Advances in the Study of Bilingualism*, pp. 93–110.

[23] Soto, V., & Hirschberg, J. (2017). Crowdsourcing universal part-of-speech tags for code-switching. ArXiv, abs/1703.08537.

[24] Singh, V., Vijay, D., Akhtar, S., & Shrivastava, M. (2018). Named entity recognition for Hindi-English code-mixed social media text. NEWS@ACL, pp. 27–35.

[25] Aguilar, G., Alghamdi, F., Soto, V., Diab, M.T., Hirschberg, J., & Solorio, T. (2018). Named entity recognition on code-switched data: overview of the CALCS 2018 shared task. CodeSwitch@ACL, pp. 138–147.

[26] Partanen, N., Lim, K., Rießler, M., & Poibeau, T. (2018). Dependency Parsing of Code-Switching Data with Cross-Lingual Feature Representations. *Proceedings of the Fourth International Workshop on Computational Linguistics of Uralic Languages*, pp. 1–17.

[27] Lawrence, C., & Riezler, S. (2016). A corpus and semantic parser for multilingual natural language querying of OpenStreetMap. NAACL, pp. 740–750.

[28] Duong, L., Afshar, H., Estival, D., Pink, G., Cohen, P., & Johnson, M. (2017). Multilingual semantic parsing and code-switching. CoNLL, pp. 379–389.

[29] Bhat, I.A., Bhat, R., Shrivastava, M., & Sharma, D. (2017). Joining hands: exploiting monolingual treebanks for parsing of code-mixing data. EACL, pp. 324–330.

[30] Sharma, A., Gupta, S., Motlani, R., Bansal, P., Shrivastava, M., Mamidi, R., & Sharma, D. (2016). Shallow parsing pipeline - Hindi-English code-mixed social media text. NAACL, pp. 1340–1345.

[31] Bharati, A., Sharma, D., Bai, L., & Sangal, R. (2008). *AnnCorra: Annotating Corpora Guidelines for POS and Chunk Annotation for Indian Languages.*

[32] Vilares, D., Alonso, M.A., & Gómez-Rodríguez, C. (2015). Sentiment analysis on monolingual, multilingual and code-switching Twitter Corpora. WASSA@EMNLP, pp. 2–8.

[33] Thelwall, M., Buckley, K., Paltoglou, G., Cai, D., & Kappas, A. (2010). Sentiment strength detection in short informal text. *Journal of the Association for Information Science and Technology* 61(12): 2544–2558.

[34] Vilares, D., Alonso, M.A., & Gómez-Rodríguez, C. (2016). *EN-ES-CS: An English-Spanish Code-Switching Twitter Corpus for Multilingual Sentiment Analysis.* LREC.

[35] Prabhu, A., Joshi, A., Shrivastava, M., & Varma, V. (2016). Towards subword level compositions for sentiment analysis of Hindi-English code mixed text. COLING, pp. 2482–2491.

[36] Lee, S.Y., & Wang, Z. (2015). Emotion in code-switching texts: corpus construction and analysis. SIGHAN@IJCNLP, pp. 91–99.

[37] Bohra, A., Vijay, D., Singh, V., Akhtar, S., & Shrivastava, M. (2018). A dataset of Hindi-English code-mixed social media text for hate speech detection. PEOPLES@NAACL-HTL, pp. 36–41.

[38] Vijay, D., Bohra, A., Singh, V., Akhtar, S., & Shrivastava, M. (2018). A dataset for detecting irony in Hindi-English code-mixed social media text. EMSASW@ESWC, pp. 38–46.

[39] Chandu, K.R., Chinnakotla, M.K., & Shrivastava, M. (2015). "Answer ka type kya he?" Learning to Classify Questions in Code-Mixed Language. *Proceedings of the 24th International Conference on World Wide Web.* May 18–22, 2015, Florence, Italy.

[40] Chandu, K.R., Loginova, E., Gupta, V., Genabith, J.V., Neumann, G., Chinnakotla, M.K., Nyberg, E., & Black, A. (2018). Code-mixed question answering challenge: crowdsourcing data and techniques. CodeSwitch@ACL, pp. 29–38.

[41] Gupta, V., Chinnakotla, M.K., & Shrivastava, M. (2018). Transliteration better than translation? Answering code-mixed questions over a knowledge base. CodeSwitch@ACL, pp. 39–50.

[42] Banerjee, S., Naskar, S., Rosso, P., & Bandyopadhyay, S. (2016). The First Cross-Script Code-Mixed Question Answering Corpus. MultiLingMine@ECIR.

[43] Banerjee, S., Moghe, N., Arora, S., & Khapra, M.M. (2018). A dataset for building code-mixed goal oriented conversation systems. COLING, pp. 3766–3780.

[44] Jayarao, P., & Srivastava, A. (2018). Intent detection for code-mix utterances in task oriented dialogue systems. *2018 International Conference on Electrical, Electronics, Communication, Computer, and Optimization Techniques (ICEECCOT)*, pp. 583–587.

[45] Arcan, M., Chakravarthi, B.R., & McCrae, J.P. (2018). *Improving Wordnets for Under-Resourced Languages Using Machine Translation.* GWC.

[46] Chakravarthi, B.R., Priyadharshini, R., Stearns, B., Jayapal, A., Sridevy, S., Arcan, M., Zarrouk, M., & McCrae, J.P. (2019). *Multilingual Multimodal Machine Translation for Dravidian Languages utilizing Phonetic Transcription*, pp. 56–63.

[47] Chakravarthi, B.R., Arcan, M., & McCrae, J.P. (2019). WordNet gloss translation for under-resourced languages using multilingual neural machine translation. MomenT@MTSummit, pp. 1–7.

[48] Dhar, M., Kumar, V., & Shrivastava, M. (2018). Enabling Code-Mixed Translation: Parallel Corpus Creation and MT Augmentation Approach. *Proceedings of the First Workshop on Linguistic Resources for Natural Language Processing*, pp. 131–140.

[49] Singh, T.D., & Solorio, T. (2017). Towards translating mixed-code comments from social media. CICLing, pp. 457–468.

[50] Menacer, M., Langlois, D., Jouvet, D., Fohr, D., Mella, O., & Smaïli, K. (2019). Machine Translation on a Parallel Code-Switched Corpus. *Canadian Conference on AI*, pp. 426–432.

[51] Khanuja, S., Dandapat, S., Sitaram, S., & Choudhury, M. (2020). A new dataset for natural language inference from code-mixed conversations. CALCS, pp. 9–16.

[52] Thara, S., & Poornachandran, P. (2018). Code-mixing: a brief survey. *2018 International Conference on Advances in Computing, Communications and Informatics (ICACCI)*, pp. 2382–2388.

[53] Çetinoglu, Ö., Schulz, S., & Vu, N.T. (2016). Challenges of computational processing of code-switching. CodeSwitch@EMNLP.

[54] Devlin, J., Chang, M., Lee, K., & Toutanova, K. (2019). *BERT: Pre-training of Deep Bidirectional Transformers for Language Understanding.* NAACL.

[55] Santy, S., Srinivasan, A., & Choudhury, M. (2021). *BERTologiCoMix: How does Code-Mixing interact with Multilingual BERT?* ADAPTNLP.

[56] Li, Y., & Fung, P. (2012). Code-switch language model with inversion constraints for mixed language speech recognition. COLING, pp. 1671–1680.

[57] Li, Y., & Fung, P. (2014). Code switch language modeling with Functional Head Constraint. *2014 IEEE International Conference on Acoustics, Speech and Signal Processing (ICASSP)*, pp. 4913–4917.

[58] Toribio, A. and Rubin, E. (1993). Code-Switching in Generative Grammar.

[59] Choudhury, M., Bali, K., Sitaram, S., & Baheti, A. (2017). *Curriculum Design for Code-switching: Experiments with Language Identification and Language Modeling with Deep Neural Networks.* ICON.

[60] Pratapa, A., Bhat, G., Choudhury, M., Sitaram, S., Dandapat, S., & Bali, K. (2018). *Language Modeling for Code-Mixing: The Role of Linguistic Theory-based Synthetic Data.* ACL.

[61] Chandu, K.R., Manzini, T., Singh, S., & Black, A. (2018). Language informed modeling of code-switched text. CodeSwitch@ACL, pp. 92–97.

[62] Gonen, H., & Goldberg, Y. (2019). *Language Modeling for Code-Switching: Evaluation, Integration of Monolingual Data, and Discriminative Training.* EMNLP/IJCNLP.

[63] Zeng, Z., Xu, H., Chong, T.Y., Siong, C.E., & Li, H. (2017). Improving N-gram language modeling for code-switching speech recognition. *2017 Asia-Pacific Signal and Information Processing Association Annual Summit and Conference (APSIPA ASC)*, pp. 1596–1601.

[64] Winata, G.I., Madotto, A., Wu, C., & Fung, P. (2018). Code-switching language modeling using syntax-aware multi-task learning. CodeSwitch@ACL.

[65] Lee, G., & Li, H. (2020). Modeling code-switch languages using bilingual parallel corpus. ACL, pp. 860–870.

[66] Lignos, C., Marcus, M. (2013). Toward web-scale analysis of codeswitching. 87th Annual Meeting of the Linguistic Society of America.

[67] Barman, U., Das, A., Wagner, J., & Foster, J. (2014). Code Mixing: a challenge for language identification in the language of social media. CodeSwitch@ EMNLP, pp. 13–23.

[68] Chittaranjan, G., Vyas, Y., Bali, K., & Choudhury, M. (2014). Word-level Language Identification using CRF: Code-switching Shared Task Report of MSR India System. CodeSwitch@EMNLP, pp. 73–79.

[69] Jhamtani, H., Bhogi, S.K., & Raychoudhury, V. (2014). *Word-level Language Identification in Bi-lingual Code-switched Texts.* PACLIC.

[70] Jurgens, D., Tsvetkov, Y., & Jurafsky, D. (2017). *Incorporating Dialectal Variability for Socially Equitable Language Identification.* ACL.

[71] Ansari, M.Z., Khan, S., Amani, T., Hamid, A., & Rizvi, S. (2020). Analysis of part of speech tags in language identification of code-mixed text. *Advances in Computing and Intelligent Systems, Springer*, pp. 417–425.

[72] Barman, U., Wagner, J., & Foster, J. (2016). Part-of-speech tagging of code-mixed social media content: pipeline, stacking and joint modelling. CodeSwitch@EMNLP, pp. 30–39.

[73] Goot, R.V., & Çetinoglu, Ö. (2021). *Lexical Normalization for Code-switched Data and its Effect on POS Tagging.* EACL.

[74] Alghamdi, F., Molina, G., Diab, M.T., Solorio, T., Hawwari, A., Soto, V., & Hirschberg, J. (2016). Part of speech tagging for code switched data. CodeSwitch@EMNLP.

[75] Ball, K., & Garrette, D. (2018). Part-of-speech tagging for code-switched, transliterated texts without explicit language identification. EMNLP.

[76] Attia, M.A., Samih, Y., El-Kahky, A., Mubarak, H., Abdelali, A., & Darwish, K. (2019). POS tagging for improving code-switching identification in Arabic. WANLP@ACL.

[77] Attia, M., Samih, Y., & Maier, W. (2018). GHHT at CALCS 2018: named entity recognition for dialectal Arabic using neural networks. CodeSwitch@ACL.

[78] Geetha, P., Chandu, K.R., & Black, A. (2018). Tackling code-switched NER: participation of CMU. CodeSwitch@ACL.

[79] Aguilar, G., Maharjan, S., López-Monroy, A.P., & Solorio, T. (2017). A multi-task approach for named entity recognition in social media data. NUT@EMNLP.

[80] Sabty, C., Sherif, A., Elmahdy, M., & Abdennadher, S. (2019). Techniques for named entity recognition on Arabic-English code-mixed data.

[81] Akbik, A., Blythe, D.A., & Vollgraf, R. (2018). Contextual string embeddings for sequence labeling. COLING, pp. 1638–1649.

[82] Winata, G.I., Wu, C., Madotto, A., & Fung, P. (2018). Bilingual character representation for efficiently addressing out-of-vocabulary words in code-switching named entity recognition. CodeSwitch@ACL.

[83] Wang, C., Cho, K., & Kiela, D. (2018). Code-switched named entity recognition with embedding attention. CodeSwitch@ACL, pp. 154–158.

[84] Winata, G.I., Lin, Z., & Fung, P. (2019). Learning multilingual meta-embeddings for code-switching named entity recognition. RepL4NLP@ACL, pp. 181–186.

[85] Winata, G.I., Lin, Z., Shin, J., Liu, Z., & Fung, P. (2019). *Hierarchical Meta-Embeddings for Code-Switching Named Entity Recognition*. EMNLP/IJCNLP.

[86] Bhat, I.A., Bhat, R., Shrivastava, M., & Sharma, D. (2018). *Universal Dependency Parsing for Hindi-English Code-Switching*. NAACL.

[87] Padmaja, S., Fatima, S., Bandu, S., Nikitha, M., & Prathyusha, K. (2020). *Sentiment Extraction from Bilingual Code Mixed Social Media Text*. Data Engineering and Communication Technology, Springer, pp. 707–714.

[88] Patra, B.G., Das, D., & Das, A. (2018). Sentiment analysis of code-mixed Indian languages: an overview of SAIL_Code-mixed shared Task @ICON-2017. ArXiv, abs/1803.06745.

[89] Mandal, S., & Das, D. (2018). Analyzing roles of classifiers and code-mixed factors for sentiment identification. ArXiv, abs/1801.02581.

[90] Lal, Y.K., Kumar, V., Dhar, M., Shrivastava, M., & Koehn, P. (2019). De-mixing sentiment from code-mixed text. ACL, pp. 371–377.

[91] Utsav, J., Kabaria, D., Vajpeyi, R., Mina, M., & Srivastava, V. (2020). Stance Detection in Hindi-English Code-Mixed Data. *Proceedings of the 7th ACM IKDD CoDS and 25th COMAD*, pp. 359–360.

[92] Rajput, K., Kapoor, R., Mathur, P., Hitkul, Kumaraguru, P., & Shah, R. (2020). *Transfer Learning for Detecting Hateful Sentiments in Code Switched Language*. Deep Learning-Based Approaches for Sentiment Analysis, Springer, pp. 159–192.

[93] Khandelwal, A., & Kumar, N. (2020). A Unified System for Aggression Identification in English Code-Mixed and Uni-Lingual Texts. *Proceedings of the 7th ACM IKDD CoDS and 25th COMAD*, pp. 55–64.

[94] Santosh, T., & Aravind, K. (2019). Hate Speech Detection in Hindi-English Code-Mixed Social Media Text. *Proceedings of the ACM India Joint International Conference on Data Science and Management of Data*, pp. 310–313.

[95] Johnson, M., Schuster, M., Le, Q.V., Krikun, M., Wu, Y., Chen, Z., Thorat, N., Viégas, F., Wattenberg, M., Corrado, G., Hughes, M., & Dean, J. (2017). *Google's Multilingual Neural Machine Translation System: Enabling Zero-Shot Translation.* Transactions of the Association for Computational Linguistics, pp. 339–351.

[96] Rao, P.R., & Devi, S. (2016). CMEE-IL: CodeMix entity extraction in Indian languages from social media Text @ FIRE 2016 - an overview. FIRE, pp. 289–295.

[97] Sequiera, R., Choudhury, M., Gupta, P., Rosso, P., Kumar, S., Banerjee, S., Naskar, S., Bandyopadhyay, S., Chittaranjan, G., Das, A., & Chakma, K. (2015). *Overview of FIRE-2015 Shared Task on Mixed Script Information Retrieval.* FIRE Workshops, pp. 19–25.

[98] Banerjee, S., Chakma, K., Naskar, S., Das, A., Rosso, P., Bandyopadhyay, S., & Choudhury, M. (2016). Overview of the mixed script information retrieval (MSIR) at FIRE-2016. FIRE, pp. 39–49.

[99] Jamatia, A, Das, A. (2016). Task report: Tool contest on pos tagging for code-mixed Indian social media (Facebook, Twitter, and WhatsApp). *Proceedings of ICON.*

[100] Khanuja, S., Dandapat, S., Srinivasan, A., Sitaram, S., & Choudhury, M. (2020). *GLUECoS: An Evaluation Benchmark for Code-Switched NLP.* ACL.

[101] Aguilar, G., Kar, S., & Solorio, T. (2020). *LinCE: A Centralized Benchmark for Linguistic Code-switching Evaluation.* LREC.

[102] Vu, N.T., & Schultz, T. (2014). Exploration of the impact of maximum entropy in recurrent neural network language models for code-switching speech. CodeSwitch@EMNLP.

[103] Vu, N.T., Lyu, D., Weiner, J., Telaar, D., Schlippe, T., Blaicher, F., Siong, C.E., Schultz, T., & Li, H. (2012). A first speech recognition system for Mandarin-English code-switch conversational speech. *2012 IEEE International Conference on Acoustics, Speech and Signal Processing (ICASSP)*, pp. 4889–4892.

Chapter 7

Legal document summarization using hybrid model

Deekshitha and Nandhini K.
Central University of Tamil Nadu

7.1 INTRODUCTION

The ever-increasing volume of information available in various domains such as healthcare, supermarkets, and legal institutions makes information management more complex and can lead to information overload. This is one of the major issues for general information users, researchers, and information managers [1]. This makes it challenging to find the required information rapidly and efficiently from various print, electronic, and online sources [1]. Hence, this triggers a technology that automatically overcomes the above problems by generating text summaries. Automatic text summarization generates a summary that includes all critical and relevant information from the original document [2,3]. So, the information arrives rapidly and does not lose the document's original intent [4].

7.1.1 Background

In recent days, automatic summarization has become a more popular and crucial task in the research field. It is gaining more importance in the current research field due to its influence over manual summarization.
To name a few [5]:

1. Summaries will reduce the reader's reading time.
2. It makes it easy to select research articles from the web.
3. Automatic summarization improves indexing efficiency.
4. Automatic summarization algorithms are pre-programmed, so they are less biased than human summarizers.
5. Question answering (QA) systems provide individualized information with the help of Personalized summaries.

It can be used in various applications such as education, business, healthcare, legal, etc. It helps users like students with learning disabilities, blind people, etc. In education, automatic summarization helps students with learning difficulties to understand the content more quickly and efficiently [6].

 DOI: 10.1201/9781003244332-7

Students can summarize the contents of the notes, the slides, and the textbook and generate a review of course content using automatic text summarization. This helps students understand information and transfer it to long-term memory [7]. Ordinary people with vision can read the documents conveniently. However, blind and visually impaired people need a braille system to read and understand documents in the form of braille text. However, reading a braille text with your fingertips is more complex and inconvenient than reading a standard text with your eyes. Automatic summarization also helps visually challenged people by translating text summaries into braille summaries [8]. It solves the inconvenience of reading standard text. In the medical field, automatic summarization helps summarize medical news, research articles, and clinical trial reports on the internet. This helps clinicians or medical students rapidly get relevant information on the web and monitor infectious disease outbreaks or other biological threats [9,10]. Automatic summarization is also found to be significant in the legal domain. It helps legal experts, professionals, and ordinary people understand lengthy legal documents. It also assists legal professionals and legal experts in finding a solution to a legal problem by referring to previous cases [11].

The automatic text summarization task is mainly categorized as extractive and abstractive summarization [3]. Extractive summarization is the process of extracting important sentences from an original document and concatenating those extracted texts to form a summary. Thus, the summary contains an exact copy of the original document with fewer sentences. In contrast, abstractive summarization generates a concise summary by reproducing the original text in a new way using advanced natural language techniques such as deep learning techniques [4]. The understanding of longer passages, information compression, and language generation make automatic text summarization a challenging task in natural language processing (NLP) since computers lack human knowledge and language capability [12,13]. In recent years, deep learning approaches have obtained very high performance on many NLP tasks such as named entity recognition (NER), parts-of-speech (POS) tagging or sentiment analysis, speech recognition, spell checking, machine translation, text summarization, and so on [14]. The journey of deep learning in automatic text summarization started with a simple recurrent neural network (RNN). Then the fewer disadvantages of RNN moved the research toward advanced deep learning models such as gated recurrent unit (GRU), long short-term memory (LSTM), transformer, etc. [15]. It is more efficient than extractive summarization because it produces a highly cohesive, information-rich, and less redundant summary [16]. This study utilized extractive and encoder-decoder based abstractive summarization models to generate legal document summaries. This is a two-stage model; in the first stage, the extractive summarization model, such as TF-IDF, was used to find essential sentences from the judgment (main document) of legal documents. In the second stage, the transformer model, an abstractive summarization model, was trained using selected legal data sets with a human-generated summary. Once the training is completed, the model generates the summary for test sets.

7.1.2 Motivation

The legal domain is a huge repository of court data generated by various legal institutions, worldwide. Republic countries like India have several legal institutions such as the Supreme Court, high courts, district courts, etc. Each court has documents related to different legal cases, such as civil, criminal, etc. Legal documents differ in size, structure, ambiguity, and vocabulary [62]. Because of this, ordinary people do not easily explore legal documents [62]. So, generating a summary of those documents is more important. Most documents contain headnotes at the beginning, which are a concise summary of a judgment written by professionals. However, this is a highly time-consuming task, and continuous human involvement is needed. So automatic summarization in the legal domain has become a more attractive field. Automatic summarization helps users quickly get meaningful sentences and refer to related cases.

7.1.3 Problem definition

In this chapter, we propose an extractive summarization method followed by an abstractive summarization, wherein an encoder-decoder-based abstractive summarization model is used for generating a final abstractive summary. The summary generated in the first stage was given as input to the next model. The extractive summarization followed by an abstractive summarization is for improving the system-generated summary of legal documents and reducing the size of the document while selecting only the sentences that are more important in generating the summary. Thus, the reduced-sized legal document was given input to the abstractive model to rewrite the meaning full summary. In this experiment, civil documents of Indian Supreme Court are collected as an experimental dataset, and initially, the TF-IDF-based extractive model is used for selecting the introductory sentences of the primary document. Secondly, the transformer model was used to build an abstractive summarization model. Finally, the Recall-Oriented Understudy for Gisting Evaluation (ROUGE) score evaluated the system-generated summary. We divided datasets into two training and testing sets. The first 80% of the data are used for training and the remaining for testing. The training set trains our model, and based on the training, the model will predict or generate the summary for the test set. The training will repeat until the epoch reaches a predefined value, i.e., 250. Then the model evaluation was done by ROUGE score.

7.1.4 Objectives and scopes

The main objective of our study is to build a model for the automatic summarization of legal judgments. We investigated the issues in the legal document summarization and listed the objectives of our present work:

1. To reduce the time and effort spent by legal professionals generating summaries.
2. To explore legal information quickly for general users.
3. To overcome the problem of the extractive model and improve the quality of the summary generated by the abstractive model.

7.1.5 Organization

Section 7.1 briefly introduces the background, motivation, problem definition, objectives, and scope of automatic text summarization.

Section 7.2 deals with a survey of automatic text summarization, which includes the discussion of different kinds of summarization methods and possible evaluation types. The survey started with feature-based methods, then moved toward frequency-based, machine learning, and graph-based techniques. Then this section ended with deep learning techniques. In the following subsection, a survey of automatic summarization in the legal domain was discussed, and finally, this chapter concluded with the need for extractive and abstractive summarization in the legal domain. Section 7.3 explains the proposed work with a block diagram and its steps. Initially, it includes a description of the dataset used in our experiment. Then, a description of the extractive method used in our model was explained with equations. Thirdly, it gives a detailed description of an abstract model named transformer with figures. The exact section includes the training mechanism of the transformer model. The last subsection of this chapter includes an evaluation section and describes the ROUGE score.

Section 7.4 discusses the evaluation of the proposed system. The results of the proposed work are compared with the existing models in the literature. The performance of our model was evaluated by comparing the auto-generated summary of the model with the human-generated summary using ROUGE.

Finally, Section 7.5 summarizes and concludes the work.

7.2 LITERATURE REVIEW

The research on automatic summarization started in the 1950s [17,24]. The Earliest instances of research on text summarization were based on text features, and this feature helps to extract salient sentences from the text. At the start, Luhn [18] proposed a word or phrase frequency method for scoring the sentence. The sentence with the highest score was selected as the summary sentence. Later, in 1969, Baxendale [19] used sentence location and word frequency as the scoring criterion for extracting salient sentences from the text.

In addition to the word frequency and position, Edmundsun (1969) [20] introduced two more features: title similarity and cue words for scoring the sentences. Later, researchers used different features for scoring the sentences. In 1989, Slaton [21] introduced the TF-IDF model, which was more effective in text summarization. Since it aims to define the importance of a keyword or phrase within a document. Nobata et al. [22] used TF-IDF values of word and sentence location, length, headline similarity, and query to score significant sentences. Varma et al. [23] added more features such as length of the words, parts-of-speech tag, familiarity of the word, named entity tag, an occurrence as a heading or subheading, and font style to score the sentences. The above-mentioned research is all about feature- and frequency-based text summarization. The above research assumes that the most important words appear more frequently in a document.

With the advent of machine learning techniques in NLP, researchers started to use these techniques in document extraction [24]. This method used features and learning algorithms to find similar sentences in the document. Based on the learning algorithms used, fields of text summarization are categorized as classification and clustering-based summarization. Classification-based text summarization considers summarization tasks as two-class classification problems. This system learns from the training samples and classifies test samples as either summary or non-summary sentences [24–26]. Initially, researchers used Naive Bayes algorithms to summarize contents by classifying sentences as summary or non-summary. A naive Bayesian algorithm is a simple classification algorithm based on the Bayes theorem. So it helps to calculate the probabilities of the words in the text. It can be trained using less training data and mislabeled data. Because of this, Naive Bayes is mainly used for text-related work. Kupiec et al. used Edmundson's (1969) features and Naive Bayes for sentence extraction [25]. Besides Edmundson features, sentence length and the presence of uppercase words are also used to train the Naive Bayes model. Using all these features, sentence probability was calculated for each sentence, and then the sentence was ranked for summary selection.

Aone et al. (1999) proposed a system named DimSum that uses summarization features such as text statistics (term frequency) and corpus statistics (inverse document frequency) to derive signature words. They found that automatic text summarization performance can be enhanced by using different combinations of features through machine learning techniques [26]. Like Naive Bayes, decision trees, support vector machines, Hidden Markov Models, and Conditional Random Fields (CRFs) are also used for automatic text summarization [24]. On the other hand, the clustering-based approach groups similar textual units such as paragraphs and sentences into multiple clusters to identify themes in standard information. Later text units are selected from clusters to form the final summary [24, 27]. Each cluster consists of a group of similar text units representing a subtopic (theme). Mainly, three factors affect sentence selection such as similarity of the sentences to the theme of the cluster (Ci), location of the sentence in the document (Li), and similarity to the first sentence in the document to which it belongs (Fi). In 2009, Kamal Sarkar used a similarity histogram-based sentence-clustering algorithm to identify multiple sub-topics (themes) from a multi-document text and select the representative sentences from the appropriate clusters to form the summary. The author divides the model into three stages: forming sentence clusters, ordering the clusters, and selecting sentences from each cluster [27]. Machine learning methods have been shown to be very effective and successful in single and multi-document summarization, specifically in classification-based summarization, where classifiers are trained to group related sentences [28].

Graph-based text summarization was introduced to the summarization field, and this method was inspired by the PageRank algorithm [29,30], which represents text as a graph. The sentences of text are represented as vertices, and the relationships between sentences as edges. Each edge is assigned a weight equal to the similarity between the sentences. Graph-based summarization can be applied to any language because it does not need language-specific linguistic processing other than sentence and word boundary detection [31]. In 2004, Erkan and Radev proposed a new approach named LexRank, computing sentence importance based on eigenvector centrality in a sentence-based graph. The TF-IDF method identifies the introductory sentences and ranks them to select meaningful sentences [32]. The research, as mentioned earlier, comes under the category of extractive summarization, and it covers significant portions of the research in NLP.

Abstractive summarization has recently gained popularity due to its ability to develop new sentences while generating a summary. It is more efficient than extractive summarization because it produces a highly cohesive, information-rich, and less redundant summary [16]. However, abstractive summarization is much more complex than extractive summarization because it is not scalable and requires natural language generation techniques for rewriting the sentences of the documents [33]. The techniques of abstractive summarization are categorized into two types: structured and semantic [34]; most are either graph- or tree-based or ontology- and rule (e.g., template)-based. These approaches work well, but fewer disadvantages of these techniques made the researchers look toward deep learning models such as Convolutional Neural Network (CNN), LSTM, Sequence-to-Sequence (Seq2Seq), transformer, etc.

Chopra et al. [35] introduced a convolutional attention-based conditional RNN model for abstractive summarization. This model was trained on the annotated version of the Gigaword corpus to generate headlines for the news articles based on the first line of each article. In this, annotations were used for tokenization and sentence separation. Li et al. [36] proposed a new framework for abstractive summarization, an encoder-decoder-based Seq2Seq model. In this study, both the generative latent structural information and the discriminative deterministic variables are jointly considered in the generation process of the abstractive summaries. In 2019, Choi et al. [37] proposed a neural attention model with document content memory (NAM-DCM) for abstractive summarization. This model was used to generate coherent sentences by reflecting on the key concepts of the original document. In this way, the document content memory update scheme effectively generated the long sequence of words required in general summaries by recognizing what contents to write or what to omit.

Chen and Nguyen (2019) proposed an automatic text summarization system for single-document summarization using a reinforcement learning algorithm and an RNN. RNN is used as a sequence model of encoder-extractor network architecture. A sentence-level selective encoding technique selects the important features and extracts the summary sentences [38]. Evaluation of the text summary is also important, like generating a summary. This method compares the system-generated summary with a human-generated summary. It was broadly classified into intrinsic and extrinsic evaluations [39]. Intrinsic methods evaluate the summarization system in itself, whereas extrinsic methods evaluate summarization based on how it affects the completion of some other task such as text classification, information retrieval, answering questions, etc. [40,41]. On the other hand, the intrinsic method evaluates the summary based on quality and informativeness. Few of the informative evaluation methods are Relative utility [41], Text grammars [42], Factoid Score [43], Basic Elements [44], Pyramid Method, ParaEval [45], DEPEVAL(summ) [46], ROUGE [47] and other popular metrics are Precision, Recall, F-measure [40], etc. The quality evaluation is based on redundancy, grammaticality, referential clarity, structure, and coherence.

In the recent past, automatic summarization was used in various domains, including education, business, health, law, etc. SMMRY, TLDR, Resoomer, Internet Abridged, Auto Highlight, ReadWrite for Google Chrome, and the online summarize tool are the few summarization tools on the web that help students as well as researchers to shorten the articles by providing key points of the articles [48]. The overwhelming amount of information in the medical field makes it difficult for clinicians to quickly find the most critical information from each patient's Electronic Health Records (EHRs) [49]. The popularity of machine learning in health care uses automatic summarization to generate relevant patient and time-specific topics from structured health record data. Few researchers used visualization interfaces [50–52], natural language [53,54], and supervised models to generate descriptions of structured time series. Automatic summarization also helps to summarize natural calamities related to news and reports [55], extract sentences from email or blogs, and provide information summary for business people, government officials, and researchers who search online through search engines to receive a summary of relevant pages found and agricultural articles [56]. Automatic summarization was also significant in the legal domain, which lessens the burden on legal professionals by summarizing legal documents. We investigated how to summarize legal data in a more accessible manner, and our findings are included in the below subsection.

7.2.1 Automatic text summarization in the legal domain

Research in legal document summarization was used in machine learning, such as by Kumar et al., who proposed a clustering-based approach to legal summarization [57]. This system clusters legal judgments based on the topics obtained from hierarchical Latent Dirichlet Allocation (hLDA) and finds a summary for each judgment within the cluster. Supervised learning plays a vital role in legal text summarization. Banno et al. used Support Vector Machines to extract essential sentences to summarize Japanese Supreme Court judgments [58].

At first, Hachey and Grover applied rhetorical analysis to legal texts in the context of English law. They labeled a sentence according to its rhetorical role in the judgment document [59]. Later, Saravanan et al. applied CRFs for rhetorical role identification from legal documents, and important sentences were generated with these identified roles [60]. The combination of keyword or key phrase matching and case-based techniques was introduced for automatic legal document summarization by Kavila et al. [11]. This system labels the sentences considered for the given judgment as positive and those that have not been considered negative. A different approach is developed by Yamada et al., where an annotation scheme is used for summarizing Japanese judgment documents. This scheme was based on the author's observations about the Japanese legal system [61]. Deepa Anand and Rupali Wagh [62] divided the summarization model into two-stage semi-supervised models: sentence label generation and summary generation. This system used a few similarity measures, such as similarity based on the overlap of important words, Similarity based on match of TF-IDF scores (STFIDF), similarity based on Rouge-L scores (SROUGE), and similarity based on sentence embeddings (SSE) to label the sentences. They used RNN and LSTM to generate a summary. Luijtgaarden et al. used a reinforcement learning model to generate an abstractive summary of legal documents [76]. There exist several tools for legal document summarization. To name a few are LetSum [63], SALOMON [64], FlexiCon [65], HAUSS [66], and so on. Existing extractive summarization models have many advantages but also some drawbacks. In the above-listed models, few are domain-independent, generate summaries for huge documents, and can translate summaries to other languages. However, the summaries generated by the above extractive summarization models were redundant [67]. Extractive summarization models suffer from the problem of coherent and cohesive summaries. Abstractive models can overcome such limitations by rewriting the sentences from an original document. In this work, we combined both the extractive and abstractive summarization models to generate a summary and investigated how this proposed model can improve the readability of a legal document summary.

7.3 METHODOLOGY

Our work aims to generate a summary of legal documents and improve the readability of extracted sentences from the document. Automatic summarization in the legal domain is mostly extractive. Researchers proposed various techniques, such as machine learning, graph-based methods, and deep learning, for legal document summarization. The significant problems with existing extractive summarization are lack of coherence and redundancy [16]. Thus, the insignificant sentences are more difficult to understand. So there is a need for an abstractive summarization model to generate a more effective and meaningful summary. In this study, we initially used the TF-IDF model to score each document sentence and rank each based on our obtained scores. This step helps reduce the length of the document by selecting weighted sentences that are more important in producing a summary. Then the result of this step, i.e., the initial extractive summary, is used as an input to the abstractive summarization model. Finally, the ROUGE score was used to evaluate the system-generated summary with a human-generated summary for evaluation measures.

7.3.1 Legal document summary

Abstractive summarization is a modern approach to the legal domain. In the second stage, the transformer-based abstractive summarization was applied to the selected civil case-based documents. First, it is essential to choose similar documents for the study. Training a model with similar documents will help the model understand the structure of the documents and try to predict the summary for unseen documents (test documents). The proposed work of our study is shown in Figure 7.1.

7.3.1.1 Data preparation

Our model's first and most important step is data preparation, making the data suitable for further processing. A distinguished writing style, very long sentences, and the massive amount of legal data make understanding legal

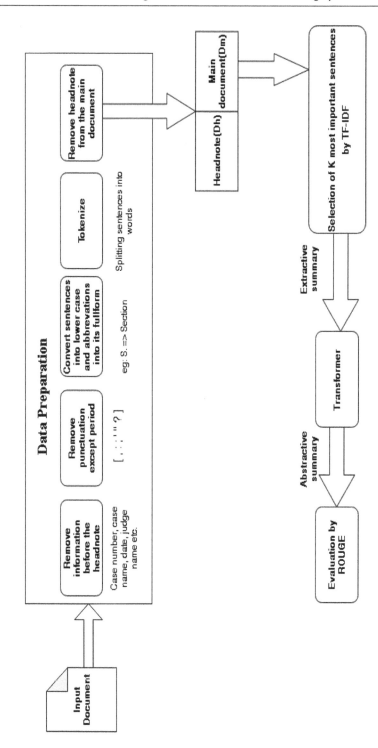

Figure 7.1 Proposed methodology.

documents more complex. We aim to make legal documents as straightforward as possible by generating a summary. Understanding the structure of the document is very basic and essential before starting the research, and the description of the dataset is given as follows:

We obtained Indian Supreme Court judgments from the site judic.nic. in. Out of all the judgments, the civil documents with headnotes were considered for our study. The headnotes section of a legal document is a concise summary of the judgment generated by legal professionals and placed at the beginning of the main document (judgment). We aim to generate a system-generated summary of the primary document. Besides the headnote and main document, each legal document contains case-related information, such as case number, date, place, judge name, etc. The section before the main document (judgment) contains a few case-related details, Act and Headnote. Each legal document contains a minimum of 3–5 pages and more than 30 pages. Here, the length of the headnote should be less than that of the primary document. The magistrate's main document or judgment for a particular case was included under the section Judgment. We removed all the information before the headnotes, such as case number, name, date, judge name, etc., from each document.

After removing the information before the headnote, we separated the contents of the headnote and the main body into two files. The main body's selected portions were input to the transformer model. The transformer model's inputs were split into train and test sets. We followed an 80%–20% train-test split, i.e., 80% of the documents were used for training and 20% for testing. In preprocessing, all punctuations except periods are removed from the document. Because periods will help identify the end of a particular sentence, the documents may contain multiple white spaces and be replaced with a single white space, and the abbreviations of a particular word are very common in legal documents. Abbreviations were replaced with their complete forms, and then every word in the sentences was tokenized and converted into lower case. After the preprocessing, the headnote was separated from the main document. In the headnote and main document, the sentences that are not necessary are still not removed. So, the next step is to find the most important sentences in the main document by calculating TF-IDF scores. Based on the scores, sentences are rearranged, and K's most important sentences are selected. The selected sentences are combined to form an initial summary in our model, which is used as an input to the transformer model. The experiment has different K values, such as 10%, 15%, 20%, and 25% of the document length. For example, the K value of 20% of the primary document means that out of 100 sentences, the 20 most similar sentences are selected as summary sentences, and the remaining are non-summary sentences. Extractive TF-IDF is explained in the below section.

7.3.1.2 Extractive TF-IDF model (generating extractive summary)

TF-IDF is a statistical measure that evaluates how relevant a word is to a document in a collection of documents [68]. It is helpful in text analysis, information retrieval, etc. TF-IDF is the most common metric for document comparison [68] and converts each sentence into a vector form. The calculation of TF and IDF is given in equations 7.1 and 7.2, respectively:

$$TF(t) = \frac{Number\ of\ times\ term\ t\ appears\ in\ a\ document}{Total\ number\ of\ terms\ in\ document} \qquad (7.1)$$

$$IDF(t) = log_e\left(\frac{Total\ Number\ of\ documents}{Number\ of\ documents\ with\ term\ t\ in\ it}\right) \qquad (7.2)$$

Hence, TF-IDF for a word can be calculated as follows:

$$TF-IDF(t) = TF(t) * IDF(t) \qquad (7.3)$$

This step helps to reduce the contents of the documents by selecting K's most important sentences. Then selected sentences are combined to form an extractive summary. Lack of cohesion and coherence, in summary, leads us to move our experiments toward the next stage, so the output of this stage is used as input to the transformer model. The description of this stage is given in Figure 7.2. Moreover, the description of the transformer model is given in the below section.

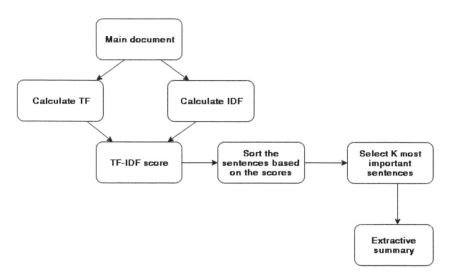

Figure 7.2 Extractive summarization model.

7.3.1.3 Abstractive summarization-transformer model

The transformer is a deep learning model proven to be an efficient neural network model for everyday NLP tasks. It was initially applied for machine translation and was recently used by OpenAI in their language models [69]. The transformer transforms a given sequence of words in a sentence into another sequence with the help of an encoder and decoder [70]. The transformer does not include RNN, but the attention mechanism combined with feed-forward networks is used to map the input and output dependencies. The attention mechanism is an essential part of the transformer, representing how critical other tokens are in the input for encoding a given token. The main components of a transformer model are discussed in the below paragraphs.

a. Word vector embeddings and positional encodings:
The transformer neural network receives an input sentence and converts it into a word vector embedding and positional encoding sequence. The word vector embeddings are a numeric representation of the text. It is necessary to convert the words to the embedding representation since a computer can only learn numbers and vectors. Word embedding is a mechanism to represent every word in an n-dimensional dense vector. Similar words will have similar vectors. On the other hand, the positional encodings are a vector that gives context based on the word's position in the original sentence. It came into existence because the exact words in different sentences may have different meanings.

Vaswani et al. [71] used sine and cosine functions for calculating positional encodings. The calculations of positional encodings are given below in equations 7.4 and 7.5:

$$PE(pos, 2i) = sin\left(\frac{pos}{10000^{\frac{2i}{d_{model}}}}\right) \tag{7.4}$$

$$PE(pos, 2i+1) = cos\left(\frac{pos}{10000^{\frac{2i}{d_{model}}}}\right) \tag{7.5}$$

with $d_{model} = 512$ (thus $i \in [0, 255]$) in the original paper [71].

The word embedding and positional encoding vectors of input and target data are combined and passed as a result to a series of encoders and decoders.

b. Encoder and decoder
The transformer consists of six encoders and six decoders [71]. Each encoder is similar in structure and has two layers: self-attention and

a feed-forward neural network. Like encoders, decoders contain self-attention, feed-forward, and an extra Masked Multi-Head Attention. Each decoder is the same as an encoder. The encoder's input will first be processed in a self-attention layer. It tells how much each word is related to every other word in the same sentence. Self-attention allows the model to look at the other words in the input sequence to better understand a particular word in the sequence [71].

c. The calculation of self-attention in transformers:

In the first step, the model creates three vectors from each of the encoder's input vectors, such as a query vector (Q), a key vector (K), and a value vector (V). These vectors are trained and updated during the training process. Next, self-attention is calculated for every word in the input sequence using equation 7.6.

$$Attention(Q, K, V) = softmax\left(\frac{QK^T}{\sqrt{d_k}}\right)V \tag{7.6}$$

Self-attention is computed multiple times simultaneously and independently in the transformer's architecture. It is therefore referred to as multi-head attention. The encoder and decoder stacks work as given below [71].

1. Initially, the word embeddings of the input sequence are passed to the first encoder.
2. Word embeddings and positional encodings will be transformed and propagated to the next encoder.
3. A set of attention vectors K and V from the last encoder output will be used by each decoder in its "encoder-decoder attention" layer, which helps the decoder focus on particular parts of the input sequence.

7.3.2 Evaluation

Lin introduced a set of matrices called ROUGE, which gives scores based on the similarity in the sequences of words between the system-generated summary and the human-generated summary. It is defined as a measure of overlapping n-grams in the system-generated and human-generated summaries [72]. ROUGE scores contain three matrices, i.e., precision, recall, and F-measure. Precision is the ratio of overlapping words to the total number of words in the system-generated summary [72]. Precision is measured as:

$$Precision = \frac{Number\ of\ overlapping\ word}{Total\ number\ of\ words\ in\ the\ system-generated\ summary} \tag{7.7}$$

Recall in the context of ROUGE means, "how much of the reference summary is the system summary recovering or capturing?"[72]. It can be calculated as:

$$Recall = \frac{Number\ of\ overlapping\ word}{Total\ number\ of\ words\ in\ the\ reference\ summary} \qquad (7.8)$$

Finally, the F-measure can be calculated using both Precision and Recall:

$$F - measure = \frac{2 * Precision * Recall}{Precision + Recall} \qquad (7.9)$$

7.4 EXPERIMENTS AND RESULTS

We experimented with 100 legal documents with headnotes using different K values. Here K is the threshold that decides the extractive summary length. The four sets of K values, i.e., 10%, 15%, 20%, and 25% of the document length, are used in our experiment. In this, the K value equals 20% of the primary document, which means out of 100 sentences, the 20 most similar sentences are selected as summary sentences, and the remaining are non-summary sentences. These K's most basic sentences were used to generate an initial summary given as input to the abstractive model to rewrite the sentences. The initial extractive summary was not cohesive. This means there is no link between the first and second sentences of the summary. Because those sentences are selected from the different sections of the main document and not in the same order as the headnote. So we added a new stage to our model with a generation of abstractive summaries. In this step, the initial summary was rewritten by the transformer model and evaluated using the ROUGE score. The first stage itself picks the most suitable or meaningful sentences. Thereby, it reduces the size of the input of the transformer model. An 80%–20% train-test split ratio was followed during the training and testing phases of the abstractive model, i.e., training samples were 80 and testing samples were 20. Table 7.1 describes the dataset and experiment, and the following sections describe the evaluation stage of our model.

Table 7.1 Dataset and experiment

Sample size	K values	Training set size	Testing set size
100	10, 15, 20, 25	80	20

7.4.1 Evaluating extractive model

We evaluated our extractive model using ROUGE. For this, Python's built-in function Rouge was used, which finds how much of the generated extractive summary is similar to the human-generated summary (headnote). The experimental results are given in Table 7.2. From Table 7.2, it is clear that as K increases, the performance of our TF-IDF model also increases. Here we considered only the precision score because precision tells how close the system-generated summary is to the reference summary [72]. As the focus is on the system-generated summary, precision gains more importance among other scores in our experiment.

7.4.2 Effects of different *K* values (summary length)

Figure 7.3 shows the results for the Rouge-1 scores with varying summary sizes. These three scores, i.e., Precision, Recall, and F-measure, were marked with blue, orange, and green colors. When the summary length is equal to 10%, recall has the highest score among other measures. Then, while increasing the summary length, the line graph gradually decreased.

Table 7.2 Extractive summarization results (ROUGE-1 score)

	K=(0.10 * document length)			K=(0.15 * document length)			K=(0.20 * document length)			K=(0.25 * document length)		
	F-measure	P	R	F-measure	P	R	F-measure	P	R	F-measure	P	R
TF-IDF	0.263	0.694	0.162	0.221	0.757	0.129	0.184	0.794	0.104	0.144	0.829	0.079

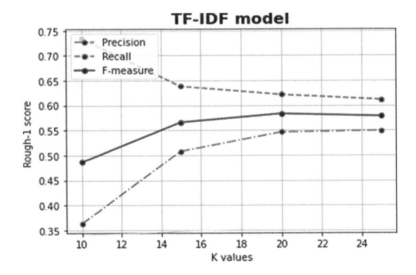

Figure 7.3 ROUGE-1 score graph for TF-IDF.

Table 7.3 Training parameters

Epoch	Batch size	Number of layers	Optimizer	Learning rate	Start token	End token
250	10	4	Adam	Customized	<go>	<end>

It means the recall value decreases when we increase the summary size. On the other hand, F-measure and Precision increase as summary length increases. The increasing Precision value for a longer summary length says the system summary is closed with a reference summary.

7.4.3 Evaluating abstractive summary model

Table 7.3 gives an overview of the training parameters used in the abstractive model. A mini-batch method with a batch size of 10 and 250 epochs was used in the training phase. The batch size is a hyperparameter defined as the number of training examples utilized in one iteration [73]. Here, batch size equals ten, meaning ten samples are trained in a single iteration. In this case, the batch size is larger than one and less than the dataset size [74]. The epoch value of 250 says the training continues until the number of epochs becomes 250. A transformer with a set of four encoders and decoders is used. It means both the encoder and decoder contain four fully connected layers. Usually, a neural network needs an optimization algorithm to optimize its weights. We used an Adam optimizer with a customized learning rate to update network weights iteratively. It is the most recommended optimization algorithm for deep learning problems [75]. In this experiment, the encoder uses an extractive summary of a document as input, and the decoder uses the headnote of trained samples as input. The starting and ending of decoder input were easily predicted by adding the <go> and <end> tokens, respectively. We fixed a decoder input's start and end positions as <go> and <end>, respectively. The model continues to predict the words or generate a summary until it reaches the end of the sequence or <end> token.

The data generated with the help of TF-IDF are used as input to the encoder model, and the headnote part of trained samples is given as input to the decoder model. We fixed the start and end positions of a decoder input as <go> and <end>. These two tokens are used to specify the starting and ending of a prediction. The model starts to train using the given samples, and training will stop until it reaches the specified epoch, i.e., 250. Then the trained model was used to predict or generate a summary for testing samples. The prediction will end when it reaches the end of the sentence token, and the ROUGE scores of the final summary are given in Table 7.4. In this table, we observe that the precision score is much higher than the Recall score, indicating that the model tends to be careful when adding new words to the summary that may not be found in the reference summary. The low recall value tells us that some words are found in the reference summary but not discussed in the summary. Figures 7.4 and 7.5 show how

Table 7.4 Final summary (ROUGE-1 scores)

	K=(0.10 * document length)			K=(0.15 * document length)			K=(0.20 * document length)			K=(0.25 * document length)		
	F-measure	P	R	F-measure	P	R	F-measure	P	R	F-measure	P	R
TF-IDF+ Transformer	0.338	0.473	0.263	0.326	0.490	0.244	0.363	0.492	0.287	0.442	0.523	0.383

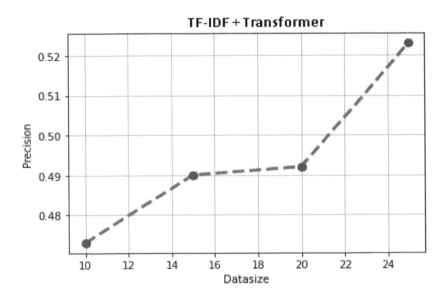

Figure 7.4 Precision graph after the final stage.

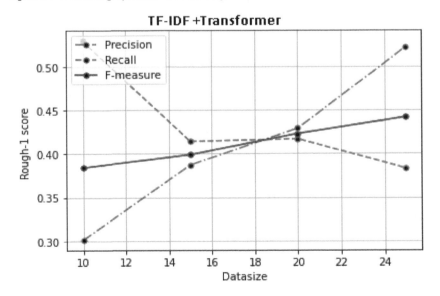

Figure 7.5 ROUGE-1 score graph after the final stage.

the model performs for different length datasets. It shows we have a higher score for extensive data among the samples. When the data size is smaller, the recall has a high score, which decreases when the data length increases.

7.4.4 Comparison with extractive summarization models

Legal document summarization has been addressed using extractive summarization methods. Encoder-decoder-based semi-supervised classification models are one of the extractive models. Those models select significant sentences from the datasets and combine them to form a summary. In this study, we used extractive and abstractive methods to improve the quality of a generated summary. We compared our model with other extractive summarization models in the literature, namely LSTM+GloVe (+Global Vectors) [62], LSTM+Word2Vec [62], Naive Bayes [62], and Random Forests [62], and the results are given in Table 7.5. The single star (*) denotes the highest precision for the given model. Results show that the extractive TF-IDF model has the highest precision score among the other models listed.

7.4.5 Comparison with abstractive model

This section compared our model with an abstractive model, also applied to the legal domain. Table 7.6 gives the experimental specifications of the compared and proposed models. They compared the model; Luijtgaarden [76] uses a sample size of 1905 legal documents and generates a summary between 40 and 150 words. Here, we experimented with 100 samples and generated a summary of 1,500 words.

Table 7.7 shows the evaluation scores (multiplied by 100) for civil case document summarization.

Table 7.5 Comparison between the proposed model and other extractive models

Type	F1-measure	Precision	Recall
Naive Bayes	0.385	0.53	0.302
Random Forests	0.345	0.382	0.315
LSTM+Glove	0.436	0.519	0.376
LSTM+Word2Vec	0.42	0.491	0.367
Extractive-TF-IDF	0.144	0.829*	0.079

Table 7.6 Experimental comparison

	Compared model (Luijtgaarden [76])	Proposed model
Generated summary size	Between 40 and 150 words	1,500
Document size	1,905	100

Table 7.7 Comparison between the proposed model and other abstractive models

Type	Recall	Precision	F1-measure
Luijtgaarden [76]	33.36	32.30	32.83
TF-IDF+Transformer	38.3*	52.3*	44.2*

Our proposed model performs better than the abstractive model in terms of Precision, Recall, and F1-Measure. So we got better results for rewriting sentences, including essential document concepts.

7.5 CONCLUSION

This study presented extractive and abstractive models with different summary lengths for legal document summarization. TF-IDF is used as an extractive model that selects a legal document's introductory sentences. This model used four different K values, such as 10%, 15%, 20%, and 25% of the length of the primary document, to evaluate the effect of summary length in generating a summary. The results show that both the extractive and extractive+abstractive models have a better precision score than the other listed models. Another important revelation from the result was that as the summary length (K) increases, the model's precision score also increases. Thus, there is a close relationship between summary length and the accuracy of the system-generated summary. However, the extractive summary generated in Stage 1 has low coherence and cohesion. A Stage-2 with an encoder-decoder-based transformer model was used to generate a final summary to overcome this limitation. The output of Stage-1 was used as an input to the Stage-2 model. This also helps reduce the size of the document by selecting only the introductory sentences of the primary document. The final results show that our TF-IDF+Transformer model outperforms the other models in terms of ROUGE, and this proposed model is more effective than the existing abstractive summarization model in the literature. It can generate suitable summaries for shorter and longer documents by reflecting the critical concepts of the document.

REFERENCES

[1] Hoq, K. M. G. (2014). Information overload: causes, consequences and remedies-a study. *Philosophy and* Progress, 49–68.
[2] Allahyari, M., Pouriyeh, S., Assefi, M., Safaei, S., Trippe, E. D., Gutierrez, J. B., & Kochut, K. (2017). Text summarization techniques: a brief survey. *arXiv preprint arXiv:1707.02268*.
[3] Gambhir, M., & Gupta, V. (2017). Recent automatic text summarization techniques: a survey. *Artificial Intelligence Review, 47*(1), 1–66.

[4] Murad, M. A. A., & Martin, T. (2007). Similarity-based estimation for document summarization using Fuzzy sets. *International Journal of Computer Science and Security*, 1(4), 1–12.

[5] Torres-Moreno, J. M. (2014). Automatic Text Summarization. https://doi.org/10.1002/9781119004752.ch1.

[6] Nandhini, K., & Balasundaram, S. R. (2013). Improving readability through extractive summarization for learners with reading difficulties. *Egyptian Informatics Journal*, 14(3), 195–204.

[7] Susar, F., & Akkaya, N. (2009). University students for using the summarizing strategies. *Procedia-Social and Behavioral Sciences*, 1(1), 2496–2499.

[8] Wan, X., & Hu, Y. (2015, July). Braillesum: A news summarization system for the blind and visually impaired people. In *Proceedings of the 53rd Annual Meeting of the Association for Computational Linguistics and the 7th International Joint Conference on Natural Language Processing (Volume 2: Short Papers)* (pp. 578–582).

[9] Afantenos, S., Karkaletsis, V., & Stamatopoulos, P. (2005). Summarization from medical documents: a survey. *Artificial Intelligence in Medicine*, 33(2), 157–177.

[10] Summerscales, R. L., Argamon, S., Bai, S., Hupert, J., & Schwartz, A. (2011, November). Automatic summarization of results from clinical trials. In *2011 IEEE International Conference on Bioinformatics and Biomedicine* (pp. 372–377). IEEE, Atlanta, GA, USA.

[11] Kavila, S. D., Puli, V., Prasada Raju, G. S. V., & Bandaru, R. (2013). An automatic legal document summarization and search using hybrid system. In *Proceedings of the International Conference on Frontiers of Intelligent Computing: Theory and Applications (FICTA)* (pp. 229–236). Springer, Berlin, Heidelberg.

[12] Liu, P. J., and Zhao, Y. (2020, June 9). PEGASUS: A state-of-the-art model for abstractive text summarization. AI Google Blog. https://ai.googleblog.com/2020/06/pegasus-stateof-art-model-for.html.

[13] Allahyari, M., Pouriyeh, S., Assefi, M., Safaei, S., Trippe, E. D., Gutierrez, J. B., & Kochut, K. (2017). Text summarization techniques: a brief survey. *arXiv preprint arXiv:1707.02268*.

[14] Gajavalli, S. H. (2021, Jan 29). *All the deep learning breakthroughs in NLP*. Analytics India. https://analyticsindiamag.com/all-the-deep-learning-breakthroughs-in-nlp/

[15] Lemberger, P. (2020, May 10). *Deep learning models for automatic summarization*. Towards Data Science.

[16] Saranyamol, C. S., & Sindhu, L. (2014). A survey on automatic text summarization. *International Journal of Computer Science and Information Technologies*, 5(6), 7889–7893.

[17] Bhatia, P. (2015). A Survey to Automatic Summarization Techniques. *International Journal of Engineering Research and General Science*, 3(5), 1045–1053.

[18] Luhn, H. P. (1958). The automatic creation of literature abstracts. *IBM Journal of Research and Development*, 2(2), 159–165.

[19] Baxendale, P. B. (1958). Machine-made index for technical literature-an experiment. *IBM Journal of Research and Development*, 2(4), 354–361.

[20] Edmundson, H. P. (1969). New methods in automatic extracting. *Journal of the ACM, 16*(2), 264–285.

[21] Salton, G; McGill, M. J. (1986). *Introduction to modern information retrieval*. McGrawHill. ISBN 978-0-07-054484-0.

[22] Nobata, C., Sekine, S., Murata, M., Uchimoto, K., Utiyama, M., & Isahara, H. (2001, March). Sentence extraction system assembling multiple evidence. In *Proceedings of the Second NTCIR Workshop on Research in Chinese & Japanese Text Retrieval and Text Summarization*. Tokyo, Japan.

[23] Pingali, J. J., & Varma, V. (2005). Sentence extraction based on single document summarization. In *Workshop on Document Summarization*.

[24] Das, D., & Martins, A. F. (2007). *A survey on automatic text summarization*. Language Technologies Institute.

[25] Kupiec, J., Pedersen, J., & Chen, F. (1995, July). A trainable document summarizer. In *Proceedings of the 18th Annual International ACM SIGIR Conference on Research and Development in Information Retrieval* (pp. 68–73).

[26] Aone, C. (1999). A trainable summarizer with knowledge acquired from robust NLP techniques. *Advances in automatic text summarization* (pp. 71–80).

[27] Sarkar, K. (2009). Sentence clustering-based summarization of multiple text documents. *TECHNIA-International Journal of Computing Science and Communication Technologies, 2*(1), 325–335.

[28] Qaroush, A., Farha, I. A., Ghanem, W., Washaha, M., & Maali, E. (2021). An efficient single document Arabic text summarization using a combination of statistical and semantic features. *Journal of King Saud University-Computer and Information Sciences, 33*(6), 677–692.

[29] Ziheng, L. (2007). *Graph-based methods for automatic text summarization*. School of Computing, National University of Singapore.

[30] Brin, S., & Page, L. (1998). The anatomy of a large-scale hypertextual web search engine. *Computer Networks and ISDN Systems, 30*(1–7), 107–117.

[31] Mihalcea, R., & Tarau, P. (2005). A language independent algorithm for single and multiple document summarization. In *Companion Volume to the Proceedings of Conference including Posters/Demos and Tutorial Abstracts*.

[32] Erkan, G., & Radev, D. R. (2004). LexRank: graph-based lexical centrality as salience in text summarization. *Journal of Artificial Intelligence Research, 22*, 457–479.

[33] Khatri, C., Singh, G., & Parikh, N. (2018). Abstractive and extractive text summarization using document context vector and recurrent neural networks. *arXiv preprint arXiv:1807.08000*.

[34] Kasture, N. R., Yargal, N., Singh, N. N., Kulkarni, N., & Mathur, V. (2014). A survey on methods of abstractive text summarization. *International Journal for Research in Emerging Science and Technology, 1*(6), 53–57.

[35] Chopra, S., Auli, M., & Rush, A. M. (2016, June). Abstractive sentence summarization with attentive recurrent neural networks. In *Proceedings of the 2016 Conference of the North American Chapter of the Association for Computational Linguistics: Human Language Technologies* (pp. 93–98).

[36] Li, P., Lam, W., Bing, L., & Wang, Z. (2017). Deep recurrent generative decoder for abstractive text summarization. *arXiv preprint arXiv:1708.00625*.

[37] Choi, Y., Kim, D., & Lee, J. H. (2018, October). Abstractive summarization by neural attention model with document content memory. In *Proceedings of the 2018 Conference on Research in Adaptive and Convergent Systems* (pp. 11–16).

[38] Chen, L., & Le Nguyen, M. (2019, October). Sentence selective neural extractive summarization with reinforcement learning. In *2019 11th International Conference on Knowledge and Systems Engineering (KSE)* (pp. 1–5). IEEE.

[39] Saziyabegum, S., & Sajja, P. S. (2017). Review on text summarization evaluation methods. *Indian Journal of Computer Science Engineering*, 8(4), 497500.

[40] Steinberger, J. (2009). Evaluation measures for text summarization. *Computing and Informatics*, 28(2), 251–275.

[41] Radev, D. R., & Tam, D. (2003, November). Summarization evaluation using relative utility. In *Proceedings of the Twelfth International Conference on Information and Knowledge Management* (pp. 508–511).

[42] Branny, E. (2007). Automatic summary evaluation based on text grammars. *Journal of Digital Information*, 8(3).

[43] Teufel, S., & Halteren, H. V. (2004). Evaluating information content by factoid analysis: human annotation and stability. In *Proceedings of the 2004 Conference on Empirical Methods in Natural Language Processing*, Barcelona, Spain: Association for Computational Linguistics (pp. 419–426)

[44] Hovy, E., Lin, C. Y., & Zhou, L. (2005, October). Evaluating DUC 2005 using basic elements. In *Proceedings of DUC* (Vol. 2005).

[45] Zhou, L., Lin, C. Y., Munteanu, D. S., & Hovy, E. (2006, June). Paraeval: Using paraphrases to evaluate summaries automatically. In *Proceedings of the Human Language Technology Conference of the NAACL, Main Conference* (pp. 447–454).

[46] Owczarzak, K. (2009, August). DEPEVAL (summ): Dependency-based evaluation for automatic summaries. In *Proceedings of the Joint Conference of the 47th Annual Meeting of the ACL and the 4th International Joint Conference on Natural Language Processing of the AFNLP* (pp. 190–198).

[47] Lin, C. Y. (2004, July). Rouge: A package for automatic evaluation of summaries. In *Text summarization branches out* (pp. 74–81).

[48] Eric. (2020, December 24). Seven Summarization Tools for Student. Control Alt Achieve. https://www.controlaltachieve.com/2017/10/summary-tools.html

[49] Gong, J. J., & Guttag, J. V. (2018, November). Learning to summarize electronic health records using cross-modality correspondences. In *Machine Learning for Healthcare Conference* (pp. 551–570). PMLR.

[50] Monroe, M., Lan, R., Lee, H., Plaisant, C., & Shneiderman, B. (2013). Temporal event sequence simplification. *IEEE Transactions on Visualization and Computer Graphics*, 19(12), 2227–2236.

[51] Plaisant, C., Milash, B., Rose, A., Widoff, S., & Shneiderman, B. (1996, April). LifeLines: Visualizing personal histories. In *Proceedings of the SIGCHI Conference on Human Factors in Computing Systems* (pp. 221–227).

[52] Hirsch, J. S., Tanenbaum, J. S., Lipsky Gorman, S., Liu, C., Schmitz, E., Hashorva, D., … & Elhadad, N. (2015). HARVEST, a longitudinal patient record summarizer. *Journal of the American Medical Informatics Association*, 22(2), 263–274.

[53] Goldstein, A., & Shahar, Y. (2016). An automated knowledge-based textual summarization system for longitudinal, multivariate clinical data. *Journal of Biomedical Informatics, 61*, 159–175.

[54] Portet, F., Reiter, E., Gatt, A., Hunter, J., Sripada, S., Freer, Y., & Sykes, C. (2009). Automatic generation of textual summaries from neonatal intensive care data. *Artificial Intelligence, 173*(7–8), 789–816.

[55] Kedzie, C., McKeown, K., & Diaz, F. (2015, July). Predicting salient updates for disaster summarization. In *Proceedings of the 53rd Annual Meeting of the Association for Computational Linguistics and the 7th International Joint Conference on Natural Language Processing (Volume 1: Long Papers)* (pp. 1608–1617).

[56] Rananavare, L. B., & Reddy, P. V. S. (2017). Automatic summarization for agriculture article. *International Journal of Applied Engineering Research, 12*(23), 13040–13048.

[57] Venkatesh, R. K. (2013). Legal documents clustering and summarization using hierarchical latent Dirichlet allocation. *IAES International Journal of Artificial Intelligence, 2*(1). http://dx.doi.org/10.11591/ij-ai.v2i1.1186.

[58] Banno, S., Mtsubara, S., & Yoshikawa, M. (2006). Identification of important parts in judgments based on machine. In *Proceedings of the 12th Annual Meeting of the Association for Natural Language Processing* (pp. 1075–1078).

[59] Grover, C., Hachey, B., & Hughson, I. (2004). The HOLJ Corpus. Supporting summarisation of legal texts. In *Proceedings of the 5th International Workshop on Linguistically Interpreted Corpora* (pp. 47–54).

[60] Saravanan, M., Ravindran, B., & Raman, S. (2008). Automatic identification of rhetorical roles using conditional random fields for legal document summarization. In *Proceedings of the Third International Joint Conference on Natural Language Processing:* Volume-I (pp. 481–488).

[61] Yamada, H., Teufel, S., & Tokunaga, T. (2017, October). Designing an annotation scheme for summarizing Japanese judgment documents. In *2017 9th International Conference on Knowledge and Systems Engineering (KSE)* (pp. 275–280). IEEE.

[62] Anand, D., & Wagh, R. (2019). Effective deep learning approaches for summarization of legal texts. *Journal of King Saud University-Computer and Information Sciences, 34*(5), 2141–2150.

[63] Farzindar, A., & Lapalme, G. (2004). LetSum, An automatic legal text summarizing. In *Legal Knowledge and Information Systems: JURIX 2004, the 17th Annual Conference* (Vol. 120, p. 11). IOS Press.

[64] Uyttendaele C., Moens M. F., & Dumortier J. (1998). Salomon: automatic abstracting of legal cases for effective access to court decisions. *Artificial Intelligence and Law, 6*(1):59–79.

[65] Smith, J., & Deedman, C. (1987). The application of expert systems technology to case-based law. In *ICAIL*, vol.87, pp 84–93

[66] Galgani, F., Compton, P., & Hoffmann, A. (2014). HAUSS: incrementally building a summarizer combining multiple techniques. *International Journal of Human-Computer Studies, 72*(7), 584–605.

[67] Polsley, S., Jhunjhunwala, P., & Huang, R. (2016, December). Casesummarizer: A system for automated summarization of legal texts. In *Proceedings of COLING 2016, the 26th International Conference on Computational Linguistics: System Demonstrations* (pp. 258–262).

[68] Umadevi, M. (2020). Document comparison based on TF-IDF metric. *International Research Journal of Engineering and Technology (IRJET)*, 7(2), 1546–1550.

[69] Giacaglia, G. (2019, October 18). How Transformers Work. Towards Data science. https://towardsdatascience.com/transformers-141e32e69591

[70] Joshi, P. (2019, June 18). *How do transformers work in NLP? A guide to the latest state-of-the-art models.* Analytics Vidhya.

[71] Vaswani, A., Shazeer, N., Parmar, N., Uszkoreit, J., Jones, L., Gomez, A. N., ... & Polosukhin, I. (2017). Attention is all you need. Advances in neural information processing systems, 30. arXiv:1706.03762.

[72] Lin, C. Y. (2004, July). Rouge: A package for automatic evaluation of summaries. In *Text summarization branches out* (pp. 74–81).

[73] Brownlee, J. (2021, January 18). Difference between a batch and an epoch in a neural network. *Machine learning mastery.* https://machinelearningmastery.com/difference-between-a-batch-and-an-epoch/

[74] Brownlee, J. (2019, January 18). How to control the stability of training neural networks with the batch size. *Machine learning mastery.*

[75] Bushaev, V. (2021, January 18). *Adam-latest trends in deep learning optimization.* Towards Data science.

[76] Luijtgaarden, N. (2019). *Automatic summarization of legal text* (Master's thesis).

Chapter 8

Concept network using network text analysis

Md Masum Billah
American International University Bangladesh (AIUB)

Dipanita Saha
Noakhali Science and Technology University

Farzana Bhuiyan
University of Chittagong

Mohammed Kaosar
Murdoch University

8.1 INTRODUCTION

A concept can be a single word or phrase representing the text's meaning. The map is the network shaped from the statements retrieving the concept network (CN) text data [1]. The statement is the overall understanding of the concepts and relations.

The CNs are extracted from articles using concept maps (CMs) and behaviorism. CN source text will be appropriate scale, and articles by different authors will emphasize other characteristics. Articles are written at different times and will also have different priorities. Meanwhile, with the development of technology, the content of the latest articles will surely be more substantial and accurate than the old ones. Two CNs on topics of different versions have many intersections of concepts and relations, so the CNs' knowledge structure is the same. However, the overall knowledge structures of the two CNs are almost the same. If one author adds some content to the old version, users can find out how much contribution the author has made to the latest version's knowledge structure through the CN. By comparing the CNs of different authors, different authors to the overall knowledge structure and the CN. Because we know the knowledge structure from the CN of the content added by different authors to the whole CN, it can also help users identify the authors' attitudes, intentions, and behaviors in the article.

DOI: 10.1201/9781003244332-8

To increase the number of open electronic texts, it is feasible to efficiently produce good tools and methods to analyze them [2]. Natural language processing (NLP) relates to reaching humans' languages, so they cannot directly understand computers. So, it is a massive provocation to explore information hidden in unstructured texts [3]. Network text analysis (NTA) is a branch of computational linguistics that uses NLP [4,5]. NTA is a method for extracting knowledge from texts and contexts and generating a network of words [4]. Such networks are semantic and can be used for different applications, for example, mental models of the authors. The NTA technique combines automated and scalable methods [4]. Automated language processing methods accelerate recycling text data and finding relevant concepts [6].

According to reference [7], a CM is a directed graph composed of concepts and relations that can be used to organize and represent knowledge structures. Nouns signify concepts or noun phrases, and relations are links between two concepts. Over the past decade, there has been a remarkable growth in the use of CMs worldwide. With the development of technologies and tools, computer-supported learning, knowledge building, and exchanging play an increasing role in online collaboration. People have more opportunities online to communicate, interact, and collaborate [8]. So, CMs can be meaningful learning tools for people. The CM can be used for collaborative writing because every author has their own opinion of one object; their article's CM will show their different views. Comparing two CMs can help other authors uncover misconceptions and knowledge conflicts between them.

NLP refers to the branch of artificial intelligence (AI) that gives machines the ability to understand and dervies its significance from human languages and combines linguistics and computer science. NLP aims to allow humans to interact with computers more easily [9]. NLP involves a representation of tasks and research areas: Syntax Analysis, Machine Translation, Semantic Analysis, Speech Recognition, Information Extraction (IE), and Discourse Analysis. It can include breaking down and separating important details from text and speech humans interact with through public social media, transferring vast quantities of freely available data to each other. NLP seems cool, yet it is a cutting-edge and complicated technology concept. It is easy to learn from a document or an article to make your algorithm understand.

Word tokenization is the task of splitting a document into document units. The document units are called tokens. A token can be a single word or a set of words. When an individual lexical unit is used as the token, it is called a unigram token. When a contiguous sequence of two lexical units is considered a token, it is called a bigram. For n items of adjacent units, the token is called an *n-gram*. Punctuation is also removed while tokenizing a document into words. For example, if we consider a sentence like: "I love science fiction," the tokens will look like as following unigram tokens:

"I," "love," "science," "fiction"

Sentence tokenization is the process of splitting a document into sentences. Punkt's sentence tokenizer [10] has been used for tokenizing sentences. NLP needs several words to build a sentence. Some words are prevalent in a sentence. These familiar words do not carry any exclusive information about the sentence. Instead, they act as glue to build a sentence. In general, these familiar words are called stop words. Stop words are considered non-meaningful or non-substantive concerning the text [11]. For example, a, an, this, that, is, was, were, etc. are stop words.

Forms for grammatical reasons such as "establish," "establishes," "establishing," and "establishment" are similar in meaning. It is convenient to search with one word and find documents containing the other words in the set [12]. Conceptually, lemmatization is almost similar to stemming. The purpose of stemming and lemmatization is to converge the conjugational forms and derivationally related words to a common base form.

For example:
am, is, are=be
girl, girls, girl's, girls'=girl
establish, establishes, establishing, establishment=establish

Stemming is the task that cuts off the ends of words to give them their base form and sometimes includes the reduction of derivational affixes. Lemmatization removes inflectional endings and creates the base or dictionary form of a word. Stemming is the process of diagnosing that some different words have the same root [13].

In most NLP, words are considered the tiniest elements with distinctive meanings [14]. Parts of speech (POS) indicate how a term is used in a sentence [15]. For example, the eight significant POS in English grammar are noun, pronoun, adjective, verb, adverb, preposition, and interjection. Classifying words into POS and labeling them is known as POS tagging [16]. Parts of speech are also known as lexical categories.

8.2 LITERATURE REVIEW

A fundamental hypothesis is that language and knowledge can be represented as a network of words and that concepts in the text represent the total content of the text [17]. The before-mentioned networks are semantic networks and can be used for different applications, for example, mental models of the authors. A cognitive model reflects the author's knowledge and understanding of a determined theme [18]. The key difference between NTA and simple keyword extraction is that it enables the extraction of both concepts and relations, and then generates a network. NTA relies on concept and relationship extraction to build semantic networks or mental models. NTA allows for the extraction of many concepts and relations and shows their ontology [19].

Concept map mining (CMM) is described as the automatic extraction or semi-automatic creation of CMs from helpful text in educational contexts.

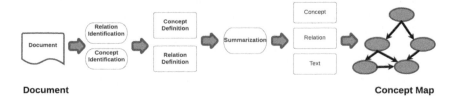

Document   **Concept Map**

Figure 8.1 Concept map mining process [20].

A CM is an accurate visual abstract of a text. The CMM process consists of recognizing the concepts in the text and the linking words that connect them. It includes concept extraction, relationship extraction, and summarization of subtasks. Concept extraction aims to identify every possible concept in the text; relationship extraction intends to find all possible connections between the previous concepts. The summarization, creating a reduced map version, summarizes the content, avoiding redundancy. Figure 8.1 is an instance of the CMM process. "Concept mapping is a type of structured conceptualization which groups can use to develop a conceptual framework which can guide evaluation or planning" [21]. A concept mapping method involving six actions and processes combined in a group brainstorming gathering. Concept Pointer Network represents copies of notable source texts as summaries that generate new conceptual words. This network leverages knowledge-based, context-aware conceptualizations to obtain an extensive collection of candidate concepts. Automatic keyword extraction (AKE) is a process that feeds several documents to extract the information that provides relevant words or other segments [22].

- Preparation of participants' selection and development for conceptualization.
- Generation of statements.
- Structuring of statements.
- Description of statements into a particular form of a CM.
- Interpretation of maps.
- Utilization of maps.

The CMM process can be implemented using NLP methods in tasks supporting IE, informal retrieval (IR), and automatic summarization [23]. Some traditional CMM methods are that they are rule-based statistical and machine learning methods. The advantages of statistical and machine learning methods are computationally efficient but not accurate enough. For more precise extraction of concepts and relations, numerical methods use dictionaries of terms for a target CM domain or linguistic tools [6]. However, there are also some problems when CMM methods process new content and context. Because dictionaries are just created for specific languages and domains, combination with linguistic tools can help solve

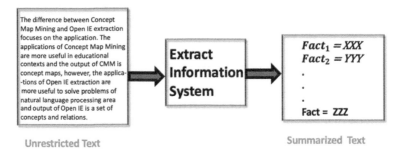

Figure 8.2 IE system for extraction of news events.

these problems, such as tokenizers, stemmers, POS taggers, parsers, and so on [23].

The process of IE automatically extracts entities and relations from unstructured textual sources. IE takes as input an available text and summarizes the text for a pre-specified topic or domain of interest. Find helpful information about the domain in the summarized text. Encode the information in a structured form suitable for populating databases [24]. The IE system is similar to a filter; texts are taken as input; and much helpful information is extracted as output. Users can reveal what they want to extract and easily manipulate the results [24]. Figure 8.2 shows a model of the IE system for extracting news events.

Typically, IE systems aim to identify and extract specific entities and relationships. Some research extends IE to many relationships and larger corpora [25–28]. However, there will be problems when target relationships are huge or impossible to have pre-specified. Open IE solves this problem by identifying relational phrases [8]. The automatic identification of relation phrases enables the extraction of arbitrary relations from sentences, obviating the restriction to a pre-specified vocabulary [28].

There are many applications of Open IE systems; Open IE systems have extensive, open-domain corpora extracted from the web, Wikipedia, and elsewhere [25,28–30]. The output of Open IE systems has been used to support tasks similar to learning sectional preferences [31]. In addition, Open IE extractions have been mapped onto existing ontologies [28]. The applications of CMM are more useful in educational contexts. The output of CMM is CMs; however, the applications of Open IE extraction are more valuable to solving problems in the NLP area, and the production of Open IE is a set of concepts and relations.

8.3 THE CONCEPT NETWORK

The CNs are extracted from wiki articles from Wikipedia and EduTechWiki platforms about CMs and behaviorism. CNs from the source text will be scaled appropriately and clearly in wiki articles. This section discussed the concept-based information retrieval of extended application fuzzy CNs.

8.3.1 Concept-based information retrieval

Semantic *concepts for* representing both documents and queries instead of keywords. This approach performs a retrieval in concept space and holds the outlook that uses high-level concepts to describe documents and queries or augment their *Bag-of-Words* (BOW) representation, which is less dependent on the specific terms used [32]. A model could find matches between different terms in the query and target documents even when the same notion is expressed, thus promoting the synonymy problem and increasing recall. If equivocal words appear in the queries, documents, and non-relevant documents retrieved with the BOW approach, correct concepts could be excluded from the results, increasing precision and easing the polysemy problem.

Concept-based methods can be defined using the following three parameters:

1. Concept representation – the "language" based on concepts. The concept-based IR approaches used *explicit* concepts representing real-life concepts that agree with human perception [33,34]. Extracting latent relations between terms or determining the probabilities of encountering terms will generate *implicit* concepts that may not adjust to any human-interpretable concept [35,36].
2. Mapping method – the method that maps natural language texts to concepts. We can use machine learning to make this mapping automatic [34], although this process usually indicates less accurate mapping. The most accurate approach would be a manual that uses a list of words to build a hand-crafted ontology of concepts that can be assigned to each [35]. However, this manual approach includes complexity and significant effort.
3. Use in IR – In this stage, concepts are used throughout the entire process in the indexing and retrieval stages [36]. Concept analysis would apply because concept-based queries increase BOW retrieval [37].

8.3.2 Concept networks and extended fuzzy concept networks

A fuzzy information retrieval method based on CNs includes nodes and directed links, where each node represents a concept or a document [38]. Two concepts are semantically related with strength μ, one connecting two distinct concept nodes, and A link associated with an accurate value μ between zero where $\mu \in [0,1]$. The extended fuzzy CNs are more common than the CNs. Fuzzy positive association, fuzzy negative association, fuzzy generalization, and fuzzy specialization are four fuzzy relationships between concepts that generate an extended fuzzy CN [39].

The fuzzy relationships between concepts and the properties of these fuzzy relationships are as follows [40].

1. Fuzzy positive association: It narrates concepts with similar fuzzy meaning (e.g., person↔individual) in some contexts.
2. Fuzzy negative association: It narrates concepts that have fuzzy complementary (e.g., men↔women), fuzzy incompatible (e.g., unemployed↔freelance), or fuzzy antonyms (e.g., tall↔short) in some contexts.
3. Fuzzy generalization: a fuzzy generalization is considered when a concept is of another concept and if it includes that concept (e.g., vehicle→bike) in a partitive sense or consists of that concept (e.g., machine→motor).
4. Fuzzy specialization: fuzzy specialization is regarded as the inverse of the fuzzy generalization relationship like (e.g., bike→vehicle) or (e.g., motor→machine). Let S be a set of concepts. Then from reference [40],
   1. "Fuzzy positive association" PA is a fuzzy relation, $PA:S{\times}S{\to}[0; 1]$, which is reflexive, symmetric, and max-*-transitive.
   2. "Fuzzy negative association" NA is a fuzzy relation, $NA:S{\times}S{\to}[0; 1]$, which is anti-reflexive, symmetric, and max-*-nontransitive.
   3. "Fuzzy generalization" GA is a fuzzy relation, $GA:S{\times}S{\to}[0; 1]$, which is anti-reflexive, antisymmetric, and max-*-transitive.
   4. "Fuzzy specialization" SA is a fuzzy relation, $SA:S{\times}S{\to}[0; 1]$, which is anti-reflexive, antisymmetric, and max-*-transitive.

8.3.3 Applications for fuzzy concept knowledge

The main feature of concept knowledge is that it is unnecessary to include the relations between all pairs of concepts to be specified because each concept is connected to all related concepts. We can calculate relations between semantically associated concepts by utilizing the inherent transitivity of the relations if they are not explicitly given. We not only take the directly linked concepts but also find related concepts by traversing one or more links. In IR, term set enlargement maps a given term to a set of equivalent terms and the inference process on the CN, which is regarded as a fuzzification. The properties of suitable applications are [41]. The knowledge graph continues the controversial number of definitions that verified the particular technical proposals where the graph of different data intended to collect and communicate knowledge has emerged. The graph of data represents the graph-based data model.

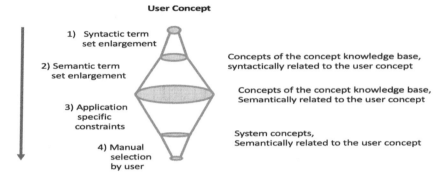

Figure 8.3 More system concepts depending on application [42].

The ontology of LCA data, costing data, and applications assigns the unfolded semantic representation (Figure 8.3).

1. The user can input the free text of its name to refer to a system concept. A terminological mismatch is anticipated to appear if the user is not familiar with that application.
2. The user concept should be mapped to one or more semantically common system concepts.
3. It is necessary to ensure a reasonable size of the set of allowed system concepts, such as a controlled vocabulary in IR.
4. It does not need the deep modeling of its domain.
5. Some structure in the set of system concepts is required, which will allow the specification of application-specific constraints for further qualification.
6. Applications with explicit negation can be used because they have negative associations in the fuzzy CN.

8.3.4 Building WikiNet: using Wikipedia as the source

Since January 2001, Wikipedia has become a large-scale source of knowledge in AI and NLP for researchers in this field. The application of Wikipedia is that it hits a middle ground between accurate, manually created, limited-coverage resources such as WordNet [43], Cyc [44], or domain-specific ontologies, dictionaries, and thesauri, and automatic, but still noisy, knowledge mined from the web [45].

Wikipedia contains much multi-faceted information: articles, links between articles, categories that group articles, infoboxes, a hierarchy that organizes the categories and articles into a large-directed network, cross-language links, etc. These different kinds of information have been used independently of each other. To produce a large-scale, multilingual, and self-contained resource, WikiNet is the result of jointly bootstrapping several information sources in Wikipedia [46]. This approach works automatically with the category and article networks, discovers relations in Wikipedia's category names, and finds numerous instances based on the category structure.

Three main steps are required for building WikiNet [46]. Firstly, to discover numerous binary relation instances, category names are formed to retrieve the categorization criterion. Secondly, information in the articles' infoboxes is used for filtering the relation instances, which were discovered in the first step. Lastly, by merging nodes that refer to the same concept, the formalized network is obtained up to this point in different languages. Moreover, add lexicalizations from a redirect, disambiguation, and cross-language links from Wikipedia for these concepts.

Like WordNet, WikiNet comprises an index of concepts covering Wikipedia articles, categories, and relationships between the concepts. To separate the lexicalization of concepts from their relationships index, this separation allows us to have a multilingual index and a language-independent relation network within the WikiNet. Various methods are used to lexicalize these concepts and extract their relationships [47].

The index involves both articles and categories. A list of integer IDs representing concepts and their lexicalizations forms the index. ID is shared between an article and its homonymous super-category. The article name, the cross-language links, anchor texts, and disambiguation links are used to collect lexicalizations. The relations connect the related concepts in the extracted index. These relations are obtained from the category network, info box relations, and relations from the article bodies. To structure the content categories in Wikipedia, users add them. Based on the type of information they encode, analysis of category names reveals different types, such as direct, partly explicit, and implicit. Infoboxes are often important enough and shared by enough entities that Wikipedia contributors use them for categorization, as they are another source of user-structured knowledge. Hyperlinks are an essential source of additional information from the article bodies [48] that highlight relevant or related concepts to the described concept. Moreover, these concept relations can successfully be used for computing semantic relatedness.

8.4 NETWORK TEXT ANALYSIS

With the progress of wireless internet and smartphone devices, data on the web is dramatically increasing, and it is the most shared content type on the web, which satisfies a large variety of user requests. To find more practical and efficient methods, many researchers in computer sciences are committed to trying to provide relevant results to meet users' demands.

Moreover, this massive amount of information also poses a severe security threat. As web data does not have semantic information, people need to spend more time understanding whether their web results are relevant or not [49]. The author's name, organizational information of the users involved, and personal information are retrieved from documents such as Microsoft Compound Document File Format [50]. The most popular document format is the Portable Document Format (PDF) to submit or share the document with others, but it might cause information leakage problems because of the diversity of privacy-related information [51]. Though detecting activities by simply extracting keywords and context words is impossible, many researchers use statistical methods such as Term Frequency (TF) or knowledge bases, such as WordNet [52–54]. Human written language is more than word frequency, so the limitation is that the precision rate is unreliable on word frequency and knowledge bases, and the results will depend on the precision of the knowledge bases even if we apply the knowledge-based approach. To understand human language, Bayes theorem [55], decision trees [56], Latent Semantic Analysis (LSA) [57], and Support Vector Machine (SVM) [58] have also been applied, and it is still a challenging task for computers to understand the text.

We can find text articles on the web using the text analysis method. WordNet hierarchies can extract context from training documents and words and build bigram data frequencies for classifying unknown text data.

8.4.1 Extracting context words from training documents

WordNet is one of the most famous knowledge bases created and maintained by the Cognitive Science Laboratory of Princeton University, and using the WordNet hierarchy, we can extract context words from the given training text articles. Semantic relationships between the words are determined by their valuable information. Concepts of hierarchy and semantic networks like synonyms, coordinate terms, hypernyms, and hyponyms can be utilized to determine the semantic distance between the words (Figure 8.4).

Using WordNet hierarchies of the concepts, WUP measurement can determine which nouns are more important than others. For example, suppose we have an article with a title and body. Using an extracted bag of words, we can obtain context word sets based on equation (1).

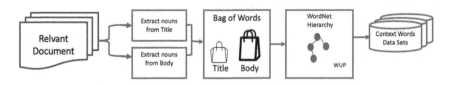

Figure 8.4 Extracting context words using WUP distance in WordNet [59].

$$sim_{wup} = 2 * depth\left(LCS\left(C1;C2\right)\right)/depth\left(C1\right) + depth\left(C2\right) \qquad (8.1)$$

where depth (C) represents the depth of concept C in the WordNet hierarchy. When two concepts share an ancestor with long depth, the value of this method goes up. According to WUP measurement, context word datasets can be obtained by calculating the average values between the bags of words.

8.4.2 Building bigram frequency for text classification

A bigram is the sequence of two adjacent elements in a string of tokens commonly used for statistical approaches in text classification. By analyzing the web pages in different fields, Google provides many n-gram data sets. These n-gram datasets can be used for classifying documents, but this approach is confined because of the huge volume of data to process [60]. The following algorithm builds the bigrams from context words for training articles.

ALGORITHM 1: ALGORITHM FOR BUILDING BIGRAM FROM CONTEXT WORDS

1: *def Bigram(str)*

2: ***str*** *← remove special characters*

3: ***splitStr*** *← space based on split in the str*

4: ***n*** *← Length of splitStr – 2 + 1*

5: ***for*** *x ← 0 to n* ***do***

6:      *vTuple ← tuple(splitStr[x:x+2])*

7:      *try:*

8:          *arr[vTuple] ← arr[vTuple] + 1*

9:      *catch:*

10:         *arr[vTuple] - 1*

11:     ***end try-catch***

12: ***end for***

In an algorithm, 'str' means the given context word sets and given data sets will be tokenized by a word. 'n' shows how many bigrams are possible in the given data set based on

$$tNgram = tWord - type + 1 \qquad (8.2)$$

The total number of possible n-grams is represented by *tNgram*, the total number of words in the given dataset is *tWord*, and the type is n-grams. According to the Google n-gram dataset, approximately 314 million, 977 million, 1.3 billion, and 1.2 billion tokens are required for the bigram, trigram, 4 gram, and 5 gram, respectively. It is best to use the bigram n-gram model to overcome size and time issues because the highest precision rate, the recall rate, and the costing time given by the 4 and 5 grams are unsuitable [61].

8.4.3 Detecting related articles by using bigrams

We need to prepare bigram datasets from related articles for training given articles corresponding to the results of queries. Two methods are used to identify related articles and compare data reliability and performances based on *bigram* weight and *Keselj*-based classification. Moreover, the procedure is shown in Figure 8.5. Then, each test bigram set and trained *bigram* set are compared based on the following equations:

$$BigramWeight = fBi_N D_i * fBi_C D_j \qquad (8.3)$$

where N represents the total number of bigrams extracted from unknown articles and K represents the total number of bigrams extracted from training articles. Their frequencies are multiplied when the unknown bigram $Bi_N D_i$ resembles the training bigram set $Bi_C D_j$. As there is a high possibility that relevant documents are more likely to share the same or similar bigrams, we must count CW (Context Weight) as we find bigram weight. If the articles describe similar subjects, then the number of the same bigrams

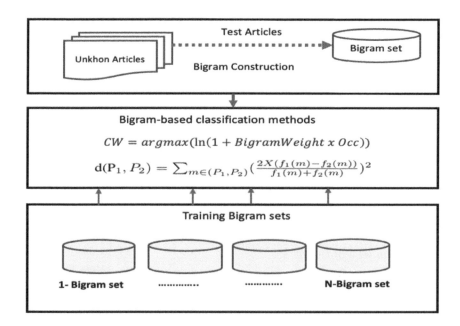

Figure 8.5 Classification steps by using n-gram-based similarities [62].

(Occ) between training and test articles will be multiplied, as shown in equation 4.

$$CW = argmax\left(ln\left(1 + BigramWeight * Occ\right)\right) \tag{8.4}$$

So, if the CW value is higher than others, then the training category will be selected for unknown articles. The most popular one for classifying documents is the n-gram-based similarity measurement named Keselj distance, based on the following equation:

$$d\left(P_1, P_2\right) = \Sigma\ m \in (P_1 \cup P_2) * \{2 * (f_1\left(m\right) - f_2\left(m\right))/f_1\left(m\right) + f_2\left(m\right)\}^2 \tag{8.5}$$

where $f_1(m)$ is the frequency of training n-gram data m and $f_2(m)$ is the frequency of unknown test n-gram data m. The training category will be selected when the *Keselj* weight is higher than others.

Using the upper steps, it is possible to detect unknown articles. It is still difficult to compare each bigram in training datasets because the size of bigrams is smaller than Google data sets. Hence, it is necessary to obtain only precise context words from the given article to ensure higher performance with less size and save costs. We can apply Wikipedia articles to extract context words to overcome the limitation of WordNet where new concepts are not defined, e.g., 'Robot.'

8.5 CONCLUSION AND FUTURE DIRECTION

A fundamentally deep syntactic analysis of the sentence still has some errors. All coreferences cannot be found using the Stanford coreference resolution annotator. The user can specify relation types of interest apriori. The relation extractor aims to identify the specified relations in the text. The mode does not need a codebook; in identifying relations, it can extract the relations interested by users and, finally, through the GraphML Export output, create a CN. Pre-specified relationships can be identified from the text.

A possible application can automatically generate CMs from students' or teachers' articles, then help students write papers or help them with their learning process. In addition, it can help the teacher to improve teaching. Another possible application can help users try to finish collaboratively writing wiki articles. Implementing a graphical user interface is also conceivable to enhance a growing range of functions and the general acceptance of tools. It is helpful for users to use it more practically and more conveniently. NTA can be a standalone application or a valuable alternative to other tools for network extraction from texts. However, it is reasonable to assume that approach will also become more accurate and scalable as dependency parsing techniques become more accurate and faster. Natural language knowledge of the content article concepts and rich ontological relationships could be developed. Meanwhile, given one or more terms of the statistical measurement of relationships, others could help rank the most likely concepts.

In the future, we plan to apply CN and NTA approaches to the audio and video speeches of some public speakers. In addition, it can help the teacher to improve teaching. Another possible application can help users finish the collaborative writing of wiki articles. Implementing a graphical user interface is also conceivable to enhance a growing range of functions and the general acceptance of tools. I believe it is helpful for users to use my approach more practically and conveniently. It is possible to integrate this approach and components of my approach as tools in an existing online workbench for knowledge extraction.

REFERENCES

[1] Carley K., & Palmquist M. (1992). Extracting, representing, and analyzing mental models. *Social Forces*, *70*(3), 601–636.
[2] Carley K. M., Columbus D., & Landwehr P. (2013). Automap User's Guide. In *CMU-ISR-13-105*.
[3] The Apache Software Foundation. UIMA Overview and SDK. https://uima. apache.org/d/uimaj-current/overview_and_setup.html
[4] Popping R. (2000). Computer-assisted text analysis. In SAGE Publications, Ltd. https://www. doi.org/10.4135/9781849208741.
[5] Diesner J., & Carley K. M. (2004). AutoMap 1.2: Extract, analyze, represent, and compare mental models from texts. In *Carnegie Mellon University. Journal Contribution*.
[6] Diesner J., & Carley K. M. (2010). Extraktion relationaler daten aus Texten. In *Handbuch Netzwerkforschung, VS Verlag fur Sozialwissenschaften*, pp. 507–521. VS Verlag für Sozialwissenschaften. https://doi.org/10.1007/978-3-531-92575-2_44
[7] Zubrinic K., Kalpic D., & Milicevic M. (2012). The automatic creation of concept maps from documents written using morphologically rich languages. *Expert Systems with Applications*, *39*(16), 12709–12718.
[8] Kimmerle J., Moskaliuk J., & Cress U. (2011). Using Wikis for learning and knowledge building: Results of an experimental study. *Educational Technology Society*, *14*(4), 138–148.
[9] Chowdhury, G. G. (2003). Natural language processing. *Annual Review of Information Science and Technology*, *37*(1), 51–89.
[10] Kiss, T., & Strunk, J. (2006). Unsupervised multilingual sentence boundary detection. *Computational Linguistics*, *32*(4), 485–525.
[11] Billah, M. M., Bhuiyan, M. N., & Akterujjaman, M. (2021). Unsupervised method of clustering and labeling of the online product based on reviews. *International Journal of Modeling, Simulation, and Scientific Computing*, *12*(2), 2150017.

[12] Manning, C. D, Raghavan, P., Schütze, H., et al. (2008). *Introduction to Information Retrieval*. Cambridge University Press, Cambridge, England.

[13] Jurafsky, D. & Martin, J. H. (2014). *Speech and Language Processing*. Vol. 3. Pearson London, London.

[14] Part of Speech. (2018). https://partofspeech.org/. Last visited: 21 February 2018.

[15] Woodward English. (2018). https://www.grammar.cl/english/parts-of-speech.htm. Last visited: 21 February 2018.

[16] Bird, S., Klein, E., & Loper, E. (2009). *Natural Language Processing with Python: Analyzing Text with the Natural Language Toolkit*. O'Reilly Media, Inc, Sebastopol, California, United States.

[17] Sowa, J. F. (1984). *Conceptual Structures: Information Processing in Mind and Machine*. Addison Wesley Longman Publishing Co., Inc, Boston, United States.

[18] Diesner, J. & Carley, K. M. (2011a). Semantic networks. In Barnett, G., editor, *Encyclopedia of Social Networking*, pp. 595–598. Sage, Thousand Oaks, California, United States.

[19] Carley, K. M. (2002). Smart agents and organizations of the future. *The Handbook of New Media*, 12, 206–220.

[20] Richardson, R. (2007). *Using Concept Maps as a Tool for Cross-Language Relevance Determination* (Doctoral dissertation, Virginia Polytechnic Institute and State University, Virginia, United States).

[21] Trochim, W. M. (1989). An introduction to concept mapping for planning and evaluation. *Evaluation and Program Planning*, 12(1), 1–16

[22] Garg, M. (2021). A survey on different dimensions for graphical keyword extraction techniques. *Artificial Intelligence Review*, 54, 4731–4770.

[23] Zubrinic, K., Kalpic, D., & Milicevic, M. (2012). The automatic creation of concept maps from documents written using morphologically rich languages. *Expert Systems with Applications*, 39(16), 12709–12718.

[24] Janevski, A. (2000). University IE: information extraction from university web pages. *University of Kentucky Master's Theses*. 217. https://uknowledge.uky.edu/gradschool_theses/217

[25] Banko, M., Cafarella, M. J., Soderland, S., Broadhead, M., & Etzioni, O. (2007, January). Open information extraction from the web. In *IJCAI* (Vol. 7, pp. 2670–2676).

[26] Mintz, M., Bills, S., Snow, R., & Jurafsky, D. (2009, August). Distant supervision for relation extraction without labeled data. In *Proceedings of the Joint Conference of the 47th Annual Meeting of the ACL and the 4th International Joint Conference on Natural Language Processing of the AFNLP*: Volume 2 (pp. 1003–1011). Association for Computational Linguistics, Suntec, Singapore.

[27] Carlson, A., Betteridge, J., Kisiel, B., Settles, B., Hruschka Jr, E. R., & Mitchell, T. M. (2010, July). Toward an architecture for never-ending language learning. In *AAAI* (Vol. 5, p. 3). AAAI Press, Atlanta, Georgia, USA.

[28] Fader, A., Soderland, S., & Etzioni, O. (2011, July). Identifying relations for open information extraction. In *Proceedings of the Conference on Empirical Methods in Natural Language Processing* (pp. 1535–1545). Association for Computational Linguistics, Edinburgh, United Kingdom.

[29] Wu, F., & Weld, D. S. (2010, July). Open information extraction using Wikipedia. In *Proceedings of the 48th Annual Meeting of the Association for Computational Linguistics* (pp. 118–127). Association for Computational Linguistics, Uppsala, Sweden.

[30] Zhu, J., Nie, Z., Liu, X., Zhang, B., & Wen, J. R. (2009, April). StatSnowball: a statistical approach to extracting entity relationships. In *Proceedings of the 18th International Conference on World Wide Web* (pp. 101–110). ACM, Madrid, Spain.

[31] Ritter, A., & Etzioni, O. (2010, July). A latent Dirichlet allocation method for selectional preferences. In *Proceedings of the 48th Annual Meeting of the Association for Computational Linguistics* (pp. 424–434). Association for Computational Linguistics, Uppsala, Sweden.

[32] Styltsvig, H. B. (2006). *Ontology-Based Information Retrieval* (Ph.D. thesis, Department of Computer Science, Roskilde University, Denmark).

[33] Voorhees, E. M. (1993). Using wordnet to disambiguate word senses for text retrieval. In *Proceedings of the 16th Annual International ACM SIGIR Conference on Research and Development in Information Retrieval*. ACM, Pittsburgh, PA, USA (pp. 171–180).

[34] Gauch, S., Madrid, J. M., Induri, S., Ravindran, D., & Chadlavada, S. (2003). *Key concept: a conceptual search engine*. Technical Report TR-8646-37, University of Kansas, United States.

[35] Deerwester, S. C., Dumais, S. T., Landauer, T. K., Furnas, G. W., & Harshman, R. A. (1990). Indexing by latent semantic analysis. *Journal of the American Society for Information Science, 41*(6), 391–407.

[36] Yi, X. and Allan, J. (2009). A comparative study of utilizing topic models for information retrieval. In *Proceedings of the 31st European Conference on IR Research (ECIR)*. Springer, Toulouse, France (pp. 29–41).

[37] Grootjen, F. & van der Weide, T. P. (2006). Conceptual query expansion. *Data and Knowledge Engineering, 56*, 174–193.

[38] Lucarella, D., & Morara, R. (1991). FIRST: Fuzzy information retrieval system, *Journal of Information Science, 17*(1), 81–91.

[39] Kracker, M. (1992). A fuzzy concept network model and its applications. In *Proceedings of the First IEEE Internet Conference on Fuzzy Systems*, IEEE, San Diego, CA, USA (pp. 761–768).

[40] Chen, S. M., & Horng, Y. J. (1999). Fuzzy query processing for document retrieval based on extended fuzzy concept networks. *IEEE Transactions on Systems, Man, Part B Cybernet, Part B (Cybernetics), 29*(1), 126–135.

[41] Kracker, M. (2002). A fuzzy concept network model and its applications. *IEEE International Conference on Fuzzy Systems*, IEEE, San Diego, CA, USA.

[42] Martin K. (1992). *A Fuzzy Concept Network Model and its Applications*. IEEE, San Diego, California.

[43] Fellbaum, C. (Ed.). (1998). *WordNet: An Electronic Lexical Database*, MIT Press, Cambridge, MA.

[44] Lenat, D. B., Guha, R., Pittman, K., Pratt, D., & Shepherd, M. (1990). Cyc: Towards programs with common sense, *Communications of the ACM, 33*(8), 30–49.

[45] Poon, H., Christensen, J., Domingos, P., Etzioni, O., Hoffmann, R., Kiddon, C., Lin, T., Ling, X., Mausam, Ritter, A., Schoenmackers, S., Soderland, S., Weld, D., Wu, F., & Zhang, C. (2010). Machine reading at the University of Washington. In *Proceedings of the NAACL HLT 2010 First International Workshop on Formalisms and Methodology for Learning by Reading*, Los Angeles, CA, USA (pp. 87–95).

[46] Vivi, N., & Michael, S. (2013). Transforming Wikipedia into a large-scale multilingual concept network, *Artificial Intelligence, 194*, 62–85

[47] Vivi, N., Michael, S., Benjamin, B., Cacilia, Z., & Anas, E. (2010). *WikiNet: A Very Large-Scale Multilingual Concept Network*, LREC, lexitron.nectec. or.th

[48] David, M., & Ian, H. (2008). Witten, an effective, low-cost measure of semantic relatedness obtained from Wikipedia links. In *Proceedings of the Workshop on Wikipedia and Artificial Intelligence: An Evolving Synergy at AAAI-08*, Chicago, IL, USA, (pp. 25–30).

[49] Hwang, M., Kim, P., & Choi, D. (2011c). Information retrieval techniques to grasp user intention in pervasive computing environment. In *Proceedings of the Innovative Mobile and Internet Services in Ubiquitous Computing (IMIS)* (pp. 186-191).

[50] Castiglione, A., Santis, A. D., & Soriente, C. (2007). Taking advantages of a disadvantage: digital forensics and steganography using document metadata. *Journal of Systems and Software, 80*(5), 750–764.

[51] Castiglione, A., Santis, A. D., & Soriente, C. (2010). Security, and privacy issues in the portable document format. *Journal of Systems and Software, 83*(10), 1813–1822.

[52] Salton, G., & Buckley, C. (1988). Term-weighting approaches in automatic text retrieval. *Journal of Information Processing and Management, 24*(5), 513–523.

[53] Hwang, M., Choi, D., & Kim, P. (2010b). A method for knowledge base enrichment using Wikipedia document information. *An International Interdisciplinary Journal, 13*(5), 1599–1612.

[54] Kong, H., Hwang, M., & Kim, P. (2005). A new methodology for merging the heterogeneous domain ontologies based on the Wordnet. In *Proceedings of the Next-Generation Web Services Practices* (pp. 22–26).

[55] Pavlov, D., Balasubramanyan, R., Dom, B., Kapur, S., & Parikh, J. (2004). Document preprocessing for naïve Bayes classification and clustering with mixture of multinomials. In *Proceedings of the Tenth ACMSIGKDD International Conference on Knowledge Discovery and Data Mining* (pp. 829–834).

[56] Lewis, D. D., & Ringuette, M. (1994). A comparison of two learning algorithms for text categorization. In *Proceedings of the Third Annual Symposium on Document Analysis and Information Retrieval*, vol. 3 (pp. 81–94).

[57] Yu, B., Xu, Z., & Li, C. (2008). Latent semantic analysis for text categorization using neural networks. *Journal of Knowledge-Based Systems, 21*(8), 900–904.

[58] Barzilay, O, & Brailovsky, V. L. (1999). On domain knowledge and feature selection using a support vector machine. *Journal of Pattern Recognition Letters, 20*(5), 475–484.

[59] Wu, Z., & Palmer, M. (1994). Verb semantics and lexical selection. In *Proceedings of the 32nd Annual Meeting of the Association for Computational Linguistics* (pp. 133–138).

[60] Choi, D., & Kim, P. (2012). Automatic image annotation using semantic text analysis. *Multidisciplinary Research and Practice for Information Systems, 7465*, 479–487.

[61] Choi, D., Hwang, M., Ko, B., & Kim, P. (2011b). Solving English questions through applying collective intelligence. *Future Information Technology, 184*, 37–46.

[62] Dongjin C., Byeongkyu K., Heesun K., & Pankoo K. (2014). Text analysis for detecting terrorism-related articles on the web. *Journal of Network and Computer Applications, 38*, 16–21.

Chapter 9

Question-answering versus machine reading comprehension

Neural machine reading comprehension using transformer models

Nisha Varghese
Christ University

M. Punithavalli
Bharathiar University

9.1 INTRODUCTION

Teaching machines to read, comprehend, and seek answers from natural language documents is not an easy task. Machine reading comprehension (MRC) has the potential to find accurate answers by asking questions based on context, and the MRC system extracts the answers through language understanding. Reading comprehension is the task of answering comprehension questions that can make sense in a textual context. MRC models facilitate the machine's ability to learn and comprehend the information from the textual contents. Neural machine reading comprehension (NMRC) has built reading comprehension systems by employing advanced deep neural networks and Transformer models. These models have shown unprecedented performance compared to other non-neural, algorithm-based, and feature-based models. The recent success of MRC is due to the contribution of large-scale datasets and the development of robust models. The accelerated MRC systems would be crucial for various applications such as search engines, virtual assistants, chatbots, and other dialogue and question-answering (QA) systems. Figure 9.1 depicts the QA and MRC systems. The significant difference between the typical QA systems and MRC is natural language understanding and comprehension. Typical QA systems utilize rule-based or keyword search algorithms to extract the answer and infer the entire sentence as an answer without any detailed language understanding. In contrast, NMRC systems use deep neural networks to scrupulously extract the correct response from the corpus with a thorough language understanding.

Researchers have identified the relevance of automatic reading comprehension systems for four decades [1]. One of the essential tasks in

DOI: 10.1201/9781003244332-9

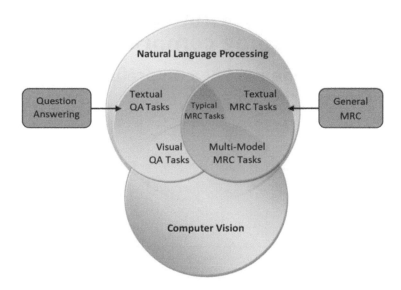

Figure 9.1 Intersecting areas of Computer Vision and Natural Language Processing and that discriminates the Question Answering and Machine Reading Comprehension and its associated Tasks.

reading comprehension is described by Hirschman, who included a dataset [2] with 60 stories of school materials. Since 2015, MRC has been at the forefront of natural language processing. The recurrent span representations (RASOR) method [3], which is an RNN model, and fixed-length representations utilize that. Hierarchical attention flow [4] for MRC leverages candidate answer options for plausible answer discrimination. The convolutional spatial attention (CSA) [5] model, which focuses on mutual information extraction from the corpus, and the attention sum reader model, which uses commonsense knowledge to deduce the accurate answer. The dual co-matching network (DMN) [6] is acquiring the semantic relationship among context-question-candidate answers in a bidirectional flow. Finally, a hybrid solution [7] exists for MRC systems without pre-given context, and the model incorporates three modules: question-candidate reader, evidence reader, and final answer prediction.

Figure 9.2 depicts the complete road map for MRC with associated components such as modules, tasks, baseline models, types of questions, recent trends, and the evaluation based on the datasets and performance evaluation metrics (PEMs). The following subsections incorporate illustrations of the components depicted in the MRC road map.

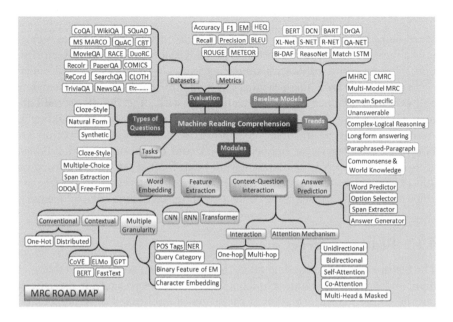

Figure 9.2 MRC Road Map – Various Tasks, Types of Questions, Baseline Models in Methodology, Performance Evaluation Metrics and various modules (stages in the complete procedure) in the Machine Reading Comprehension.

9.2 ARCHITECTURE OF MACHINE READING COMPREHENSION

MRC consists of four modules to extract an accurate answer from the context, as shown in Figure 9.3. The MRC systems can extract an accurate answer from the input context and question. The MRC systems take the input context C and question Q together; the MRC model will provide the plausible answer A to question Q by the function F, such that $A = f(Q,C)$. Word Embedding, Feature Extraction, Context-Question Interaction, and Answer Prediction are the modules of MRC.

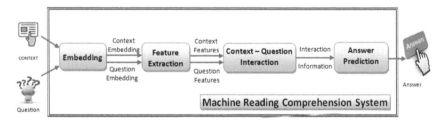

Figure 9.3 Machine Reading Comprehension Architecture with all the stages (Embedding, Feature Extraction, Context Question Interaction and Answer Prediction) in between Question-Context to Answer extraction.

9.2.1 Word embedding

Word embedding converts the text tokens to fixed-size vectors; these vectors hold a numeric value for each token. This conversion is also termed vectorization. Consequently, words with similar meanings are represented in the adjacent vector space area. These vectors can hold information about the semantic and syntactic similarity and the relationship with other words in the context. The semantics of a word lie in the context of the words surrounding it. According to the type and ability to capture the information, these vectors are bifurcated into sparse, or frequency-based, and dense, or prediction-based, vectors. The sparse vector representation holds limited semantic information and exploits huge memory, but the dense vector representation models can efficiently preserve the semantic contents without sacrificing colossal memory.

9.2.2 Feature extraction

The feature extraction module takes the question and context embedding as input and extracts the context and question features separately. It also checks out the contextual information at the sentence level based on syntactic and semantic information encoded by the embedding module. The extracted features from this module transfer to the Context-Question Interaction module.

9.2.3 Context-question interaction

The correlation between the context and the question plays a significant role in predicting the answer. With such information, the machine can determine which parts of the context are more important to answering the question. To achieve that goal, the attention mechanism, unidirectional or bidirectional, is widely used in this module to emphasize parts of the context relevant to the query. To sufficiently extract their correlation, the interaction between the context and the question sometimes involves multiple hops, simulating human comprehension's rereading process.

9.2.4 Answer prediction

Answer prediction predicts the final answer based on the information accumulated from previous modules. For cloze tests, the output of this module is a word or entity in the original context, while the multiple-choice task requires selecting the correct answer from candidate answers. Span extraction extracts a subsequence of the given context as the answer in this answer prediction module.

All the modules and methods associated with each module are depicted in Figure 9.4. Embedding can use conventional and contextual methods for word vectorization. The contextual methods can preserve more semantic

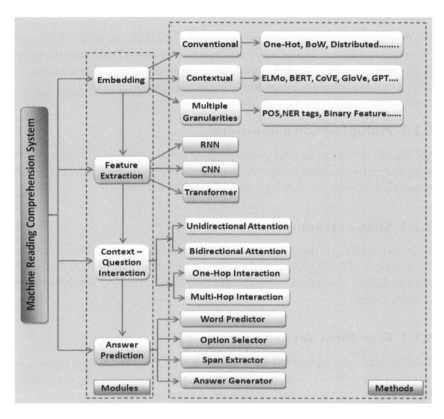

Figure 9.4 Machine Reading Comprehension Modules, Methods and Techniques using for each module to attain each task of a module.

information than the conventional methods. The context-question interactions can use attention mechanisms and one-hop or multi-hop interactions. The attention mechanism is the technique of giving attention to a specific word to capture the semantic information, and the one-hop and multi-hop methods that learn the information are similar to human learning by reading a specific context. The multi-hop methods treat the context more than once to acquire the relevant information from the context. Finally, the answer prediction module infers the exact answer based on the MRC tasks.

9.3 MACHINE READING COMPREHENSION TASKS AND CLASSIFICATION

The MRC tasks are mainly bifurcated into four types based on the type and way of extracting the answers: cloze tests, multiple-choice questions (MCQ), span extraction, and free answering.

9.3.1 Cloze tests

Cloze tests or gap-filling tests are like fill-in-the-blank questions. This type of question contains one or more blank spaces or gaps. To answer the questions, MRC systems should emphasize a comprehensive understanding of context and vocabulary usage to fill in the blanks with the missing items.

9.3.2 Multiple-choice questions

MCQs are similar to factoid questions in that they contain a list of answers with one or more correct answers. A proper understanding of the context is required to choose the correct answer from the options.

9.3.3 Span extraction

Span extraction reading comprehension is much more natural than the cloze-style and MCQ MRC tasks; specific words or entities are not sufficient to answer the questions, and there may be no proper candidate answer in some cases.

9.3.4 Free-form answering

Free-form answering delivers more accurate and flexible answers from various data repositories, which reduces the shortcomings of the tasks mentioned above due to insufficient passage and incompatible choices.

9.3.5 Attribute-based classification

The traditional task-based classifications are inadequate due to the advent of recent trends, such as multi-model, conversational, and multi-hop MRC models, and the MRC that needs commonsense, world knowledge, and logical reasoning. Figure 9.5 depicts the classification of MRC tasks based on the source of answers and types of corpora, questions, and answers.

9.4 DATASETS

The recent progress in the MRC is attributed to the development of robust large-scale datasets. The datasets with the corpus source, reference, MRC type, and links to the datasets and leaderboard are enlisted in Table 9.1, including the benchmark datasets from 2015 to 2020.

The document type represents the document or paragraph and the type of MRC, such as textual or multi-model. The multi-model datasets contain images or videos, such as Comics, MovieQA, and RecipieQA. The sources of the datasets are mainly taken from News, Wikipedia, Movies, Case Reports, Hotel Comments, Search Engines, Websites, URLs, Recipes,

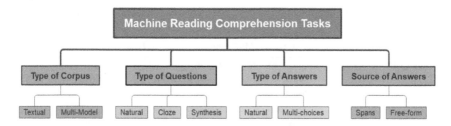

Figure 9.5 Attribute-Based Classification according to the type of corpus, questions, answers and source of answers and their associated sub-classifications.

Table 9.1 Classification of MRC bench marked datasets (2015–2020)

Dataset and reference	Source of document	Document type	MRC type*	Links & details**
CNN/Daily Mail [17]	News	Document	LSD	https://github.com/deepmind/rc-data SC = 92579/219506, SQ = 387420/997467, ST = 98.16%/88.17%
Curated TREC [18]	Factoid Stories	Paragraph	ODQA	https://github.com/brmson/ dataset-factoid-curated
WikiQA [19]	Wikipedia	Paragraph	UQ	https://www.microsoft.com/en-us/ download/details.aspx?id=52419 SC = 29258, SQ = 3047, ST = 69.51%
WikiMovies [20]	Movies	Document	RWK, DS	http://www.thespermwhale.com/ jaseweston/babi/movieqa.tar.gz https://research.fb.com/downloads/ babi/, SQ = 116K, ST = 81.33%
SQuAD 1.1 [14]	Wikipedia	Paragraph	CR	https://rajpurkar.github.io/ SQuAD-explorer/ SC = 536, SQ = 107702, ST = 81.33%
Who-did-What [21]	News	Document	PP, CR	https://tticnlp.github.io/who_did_what/ index.html, SC = 147786, SQ = 147786, ST = 86.47%
MS MARCO [22]	Bing	Paragraph	UQ, MH-MRC	http://www.msmarco.org/
NewsQA [23]	News	Document	CR	https://www.microsoft.com/en-us/ research/project/newsqa-dataset/ SC = 1010916, SQ = 119K, ST = 89.92%
Facebook CBT [24]	Factiod Stories	Paragraph	CR	https://research.fb.com/downloads/ babi/SC – 108
BookTest [25]	Factiod Stories	Paragraph	LSD	https://ibm.biz/booktest-v1, SC = 14062
Google MC-AFP [26]	Gigaword	Paragraph	CR	https://github.com/google/mcafp, SC = 1742618

(Continued)

Table 9.1 (Continued) Classification of MRC bench marked datasets (2015–2020)

Dataset and reference	Source of document	Document type	MRC type*	Links & details**
MovieQA [27]	Movies	Paragraph with images & videos	MM-MRC	http://movieqa.cs.toronto.edu/home/, SC = 548, SQ = 21406, ST = 66.18%
TriviaQA-Web [28]	Bing	Paragraph	CR	http://nlp.cs.washington.edu/triviaqa/ SC = 662659, SQ = 95956, ST = 79.72%
TriviaQA-Wiki [28]	Bing	Paragraph	CR	http://nlp.cs.washington.edu/triviaqa/ SC = 138538, SQ = 77582, ST = 79.77%
RACE [29]	English Exam	Document	CR	www.cs.cmu.edu/glai11/data/race/ https://github.com/qizhex/RACE_AR_ baselines, SC = 27933, SQ = 97687, ST = 89.95%
Quasar-S [30]	Stack overflow	Paragraph	ODQA	https://github.com/bdhingra/quasar, SC = 37362, SQ = 37362, ST = 83.10%
Quasar-T [30]	Stack overflow	Paragraph	ODQA	https://github.com/bdhingra/quasar, SC = 43012, SQ = 43013, ST = 86.50%
SearchQA [31]	J! Archive & Google	Paragraph & URL	ODQA	https://github.com/nyu-dl/SearchQA, SC = 140461, SQ = 140461, ST = 71.07%
NarrativeQA [32]	Movies	Document	MH-MRC, CR	https://github.com/deepmind/ narrativeqa SC = 1572, SQ = 46765, ST = 70.02%
SciQ [33]	School Curricula	Paragraph	DS	http://data.allenai.org/sciq/, SQ = 13679, ST = 85.38%
Qangaroo-WikiHop [33]	Scientific paper	Paragraph	MH-MRC	http://qangaroo.cs.ucl.ac.uk/index.html SC = 81318, SQ = 51318, ST = 85.23%
TQA [34]	School science Curricula	Paragraph with images	MM-MRC	http://vuchallenge.org/tqa.html, https:// openaccess.thecvf.com/content_ cvpr_2017/html/Kembhavi_Are_You_ Smarter_CVPR_2017_paper.html SC = 1076, SQ = 26260, ST = 57.71%
COMICS-Coherence [35]	Comics	Paragraph with images	MM-MRC	https://obj.umiacs.umd.edu/comics/ index.html
QuAC [36]	Wikipedia	Document	UQ, CMRC	http://quac.ai, SC = 8845, SQ = 98275, ST = 85.03%
CoQA [37]	Jeopardy	Paragraph	UQ, CMRC	https://stanfordnlp.github.io/coqa/, SC = 8399, SQ = 127K, ST = 86.61%
SQuAD 2.0 [15]	Wikipedia	Paragraph	UQ	https://rajpurkar.github.io/ SQuAD-explorer/, SC = 505, Q = 151054, ST = 86.27%

(Continued)

Table 9.1 (Continued) Classification of MRC bench marked datasets (2015–2020)

Dataset and reference	Source of document	Document type	MRC type*	Links & details**
HotpotQA- Distractor Fullwiki [38]	Wikipedia	Multi-Paragraph	MH-MRC, CR	https://hotpotqa.github.io/, SQ = 105374, ST = 85.95%
DuoRC-Self DuoRC-	Movies	Paragraph	RWK	https://duorc.github.io/, SQ = 84K, ST = 70.00%
Paraphrase [39]	Movies	Paragraph	PP, CR, UQ	https://duorc.github.io/, SQ = 100316, ST = 70.00%
CLOTH [40]	English Exam	Document	CR	https://github.com/qizhex/Large-scale-Cloze-Test-Dataset-Created-by-Teachers, SQ = 99433, ST = 77.29%
ReCoRD [41]	News	Paragraph	RWK	https://sheng-z.github.io/ReCoRD-explorer/, SC = 80121, SQ = 120730, ST = 83.43%
CliCR [42]	BMJ Case Reports	Paragraph	DS	https://github.com/clips/clicr, SQ = 104919, ST = 87.06
ReviewQA [43]	Hotel Comments	Paragraph	DS	https://github.com/qgrail/ReviewQA/, SC = 100000, SQ = 587492, ST = 89.99%
ARC-Challenge Set [44]	School science Curricula	Paragraph	CR	http://data.allenai.org/arc/, SQ = 2590, ST = 43.20%
ARC-Easy Set [44]	School science Curricula	Paragraph	RWK	http://data.allenai.org/arc/, SQ = 5197, ST = 43.31%
OpenBook QA [45]	School science Curricula	Paragraph	RWK	https://leaderboard.allenai.org/open_book_qa/, SQ = 5957, ST = 83.21%
SciTail [46]	School science Curricula	Paragraph	DS	http://data.allenai.org/scitail/ https://leaderboard.allenai.org/scitail/submissions/public, SC = 27026, SQ = 1834, ST = 84.08%
MultiRC [47]	News & Web pages	Multi-Sentence	MH-MRC	http://cogcomp.org/multirc\
RecipeQA-Cloze, RecipeQA-Coherence [48]	Recipes	Paragraph with images	MM-MRC	http://hucvl.github.io/recipeqa, SC=19779, SQ = 36K, ST = 80.62%
PaperQA-Title [49]	Scientific Papers	Paragraph	DS	http://dmis.korea.ac.kr/downloads?id=PaperQA, SQ = 84803, ST = 91.15%
PaperQA-Last [49]	Scientific Papers	Paragraph	DS	http://dmis.korea.ac.kr/downloads?id=PaperQA, SQ = 80,118, ST = 89.62%

(Continued)

Table 9.1 (Continued) Classification of MRC bench marked datasets (2015–2020)

Dataset and reference	Source of document	Document type	MRC type*	Links & details**
PaperQA [49]	Scientific Papers	Paragraph	DS	http://bit.ly/PaperQA
MCScript [50]	Narrative Text	Paragraph	RWK	https://competitions.codalab.org/ competitions/17184, SQ = 13,939, ST = 69.81%
ProPara [51]	Process Paragraph	Paragraph	CR	http://data.allenai.org/propara https:// leaderboard.allenai.org/propara/ submissions/public, SQ = 488, ST = 80.12%
Natural Questions-Short, Long [52]	Wikipedia	Paragraph	UQ	https://github.com/google-research-datasets/natural-questions, SC = 323045, SQ = 323045, ST = 95.15%
DREAM [53]	English Exam	Dialogues	RWK, CMRC	https://dataset.org/dream/, SC = 6444, SQ = 10197, ST = 59.98%
ShARC [54]	Government Websites	Paragraph	CMRC	https://sharc-data.github.ioSC=32436 SQ = 948, ST = 66.24%
Common Sense QA [55]	Narrative Test	Paragraph	RWK	https://www.tau-nlp.org/common senseqa, SQ = 12102, ST = 80.49%
DROP [56]	Wikipedia	Paragraph	CR	https://leaderboard.allenai.org/drop https://s3-us-west-2.amazonaws.com / allennlp/datasets/drop/drop_dataset. zip SC = 6735, SQ = 96567, ST = 80.16%
LogiQA [57]	Expert questions	Paragraph	LR	https://github.com/lgw863/LogiQA-dataset, SC = 8,678, ST = 80.00%
CovidQA [58]	Scientific Papers	Paragraph	DS	https://github.com/deepset-ai/ COVID-QA, SQ = 2019 (QA Pair)
ReClor [59]	Standard graduate admission exams	Paragraph	LR	https://docs.google.com/forms/u/0/d/e/ 1FAIpQLSe56wq5xIEGbDsgRN6P6l Hr34jfv182AMuilzjJXmVYCg39oA/ formResponse, SC = 6,139, ST = 75.76%

* Symbols indicating the types of MRC are: Large-Scale Dataset (LSD), Open Domain Question-Answering (ODQA), Unanswerable Questions (UQ), Require World Knowledge (RWK), Domain-Specific (DS), Complex Reasoning (CR), Paraphrased Paragraph (PP), Conversational (CMRC), Multi-model (MM-MRC), Multi-hop (MH-MRC), and Logical Reasoning (LR).
** Symbols indicating the details of MRC are Size of the Corpus (SC), Size of the Question (SQ), and Size of Training data in percentage (ST).

Comics, Books, Narrative Texts, Factoid Stories, Scientific Articles, Exams, School Curriculum Materials, and much more. The textual and multi-model with attribute-based classifications with question type, answer type, corpus type, and answer source with representative data sets are depicted in the sunburst chart (Figure 9.6) of MRC.

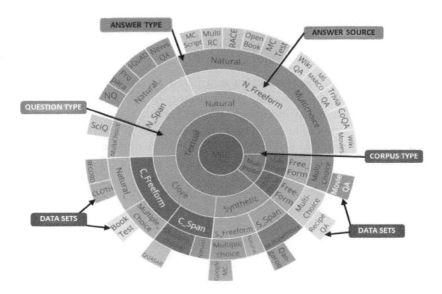

Figure 9.6 Sunburst Chart of the MRC, each circle from the Centre represents the type of corpus, question type, answer source and type of answers, the outer circle represents the representative datasets for the corresponding classification.

9.5 PERFORMANCE EVALUATION METRICS

PEMs for MRC evaluate the quality and efficiency of an MRC system or a model. PEM measures the exactness of the MRC tasks using corresponding evaluation metrics. All metrics are not appropriate for all tasks, so they have been chosen for a dataset based on their characteristics. Accuracy is the most common metric for MRC models. The following are the measures for MRC tasks:

9.5.1 Accuracy

Accuracy represents the percentage of questions an MRC model can answer correctly. The ratio between the total number of questions (N) and the number of questions the model adequately answers (M). It can be measured by equation (9.1).

$$\text{Accuracy} = \frac{M}{N} \tag{9.1}$$

9.5.2 Exact match

The exact match (EM) is the percentage of questions in which the system-generated answer exactly matches the correct answer. To recapitulate, the answer generated by the model and the correct answer should be equal. The EM can be calculated by equation (9.2).

$$Exact\ Match = \frac{M}{N} \tag{9.2}$$

9.5.3 Precision and recall

The precision and recall are calculated based on the token and question levels. Precision is the closeness of two or more measurements, or the ratio of correctly predicted positive observations to the total number of predicted positive observations. The recall is the ratio between the correctly predicted positive observations and all observations in the existing class. The precision and recall can be estimated using equations (9.3 and 9.4), respectively.

$$Precision = \frac{TP}{TP + FP} \tag{9.3}$$

$$Recall = \frac{TP}{TP + FN} \tag{9.4}$$

9.5.4 F1 score

F1 Score or balanced F-score or F-measure represents the balance between precision and recall. F1 Score is the weighted average of Precision and Recall, and in MRC, the F1 Score also has Token-level F1 and Question-level F1, similar to precision and recall. Token-level and question-level F1 for a single question can be estimated by equations (9.5 and 9.6).

$$F1_{TS} = \frac{2 \times Precision_{TS} \times Recall_{TS}}{Precision_{TS} + Recall_{TS}} \tag{9.5}$$

$$F1_{Q} = \frac{2 \times Precision_{Q} \times Recall_{Q}}{Precision_{Q} + Recall_{Q}} \tag{9.6}$$

where $F1_{TS}$ = token-level F1, $Precision_{TS}$ = token-level Precision, $Recall_{TS}$ = token-level Recall, $F1_{Q}$ = question-level F1, $Precision_{Q}$ = question-level precision, and $Recall_{Q}$ = question-level Recall for a task.

9.5.6 ROUGE (recall-oriented understudy for gisting evaluation)

Recall-oriented understudy for gisting evaluation (ROUGE) measure is used to evaluate the performance of text summarization, machine translation, MRC models, and other NLP tasks. The ROUGE measure can be estimated using equation (9.7).

$$ROUGE-N = \frac{\sum_{S \in \{RS\}} \sum_{gram_n \in S.} Count_{match}(gram_n)}{\sum_{S \in \{RS\}} \sum_{gram_n \in S} Count(gram_n)} \tag{9.7}$$

where RS is Reference Summaries, n is the n-gram length, $Count(gram_n)$ is the total number of the n-grams in the candidate text and generated text.

9.5.7 BLEU (bilingual evaluation understudy)

Bilingual evaluation understudy (BLEU) is a corpus- or text-oriented metric that measures the candidates' relatedness to the reference texts. The result is close to one of the texts with similar semantics. The BLEU can be measured by equation (9.8).

$$BLEU = min\left(1, \exp\left(1 - \frac{reference - length}{output - length}\right)\right)\left(\prod_{i=1}^{4} Precision_i\right)^{\frac{1}{4}} \qquad (9.8)$$

9.5.8 METEOR (Metric for Evaluation of Translation with Explicit ORdering)

Metric for Evaluation of Translation with Explicit ORdering (METEOR) combines Recall and Accuracy to evaluate the model's performance, and the Score is determined by the harmonic mean of precision and recall. METEOR can be measured by equation (9.9) in connection with F_{mean} and penalty, which can be estimated using equations (9.10) and (9.11). The penalty is calculated with the number of matched words and chunks, where m is the total number of matched words, ch is the number of chunks, and α, β, and γ are the parameters.

$$Meteor = F_{mean} \times (1 - Penality) \qquad (9.9)$$

$$F_{mean} = \frac{Precision \times Recall}{\propto \times Precision + (1 - \propto) \times Recall} \qquad (9.10)$$

$$Penality = \gamma \times \left(\frac{ch}{m}\right)^{\beta} \qquad (9.11)$$

9.5.9 HEQ (human equivalence score)

The human equivalence score (HEQ) is a recently developed metric for judging whether the system's output is equivalent to the output of an ordinary person. HEQ is primarily suitable for Conversational MRC due to the multiple valid answers. The HEQ score can be measured by the equation (9.12) with the number of questions N, and M is the number of questions that reach the token-level F1 of human judgment.

$$HEQ = \frac{M}{N} \qquad (9.12)$$

9.6 TRANSFORMER AND BERT

The transformer is an exceptional innovation in Deep Learning and Natural Language Processing, introduced by Ashish Vaswani et al. (2017). The transformer has shown outstanding performance on various NLP tasks. Transfer learning and parallelization are notable features of the transformer. Transfer learning is a combined process of pre-training and fine-tuning. The pre-trained data can be later used for similar tasks. The transformer has six-layered encoder-decoder stacks. The pre-training facilitates the models for fast development, less data, and better results. Parallelization is the execution of neural networks in parallel. Attention mechanism, multi-head attention, positional encoding (PE), linear layer (fully connected neural network), and softmax layer are the components of a transformer, as shown in Figure 9.7. The attention mechanism focuses on the most relevant word at each step and provides information for each input word. Multi-Headed Self Attention learns contextual relations between words or sub-words in text. The PE gives information about the order of tokens in the input sequence. The linear layer is a Fully Connected Neural Network and converts the outputs from the decoding layer to words.

Bidirectional Encoder Representations from Transformers (BERT) is one of the incredible Transformer models by Google [8] based on the Transformers with a multi-layered encoder-decoder stack. BERT is the first

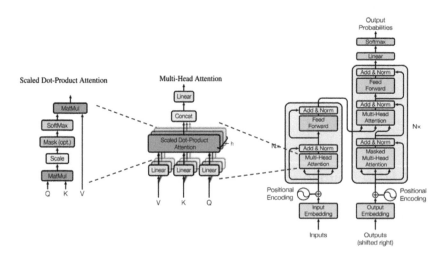

Figure 9.7 Architechture of Transformers, Scaled Dot product attention, Multihead attention, and Positional Encoding [16].

fully bidirectional model designed to pre-train from unlabeled text with an encoder-decoder structure. BERT is dramatically accelerated by all NLP and NLU Tasks through unsupervised pre-training and supervised fine-tuning procedures. The BERT base model has 12-layers of encoder-decoder stacks with roughly 110 million weights (BERT Large 24 Layers and 345M of Parameters). The reason behind the impressive performance of the BERT is the enormous size of the parameters. A considerable model with a vast corpus requires a lot of GPU memory for execution. Consequently, developing and pre-training a model is both time- and memory-consuming and cost-effective.

9.6.1 BERT-based models

This subsection included the BERT-based models such as DistilBERT, ALBERT, RoBERTa, ELECTRA, and Longformer for reading comprehension. These models are Transformer-BERT-based models with variations in the number of encoder-decoder stacks and parameters. Furthermore, some models have added features to improve their performance.

9.6.1.1 DistilBERT

DistilBERT [9] is a distilled version, a smaller language model, and faster than BERT with reduced parameters. This model has only 66M parameters, which are 40% fewer than the BERT base model, performs 60% faster, and provides accuracy to the extent of BERT. DistilBERT can execute on a CPU with a lower computation capability, and the model is pre-trained on the raw texts with an automatic labeling process. The pre-training of the DistilBERT includes some objectives such as Distillation loss, masked language modeling, and Cosine embedding loss.

9.6.1.2 ALBERT

The ALBERT [10] is A Lite BERT model for Self-supervised Learning of Language Representations with only 12 million parameters. The model introduces two parameter-reduction techniques to reduce memory consumption and accelerate training speed. ALBERT reduced approximately 89% of parameters than BERT by splitting the embedding matrix into two sub-matrices and utilizing the repeated layer split.

9.6.1.3 RoBERTa

RoBERTa (Robustly Optimized BERT Pre-training Approach) [11] was developed in collaboration with Facebook and Washington University. RoBERTa pre-trained on an enormous amount of English data, reducing the pre-training time by using a self-supervision method. The model was generated by modifying the critical hyperparameters of BERT that involve dynamically changing the masking pattern, training with sizable mini-batches and learning rates, and eliminating the Next Sentence Prediction (NSP), one of the primary features in the BERT model.

9.6.1.4 ELECTRA

ELECTRA is the abbreviation for Efficiently Learning an Encoder that classifies token replacements accurately; the model [12] introduced a novel pre-training approach called Replaced Token Detection (RTD), and the model trains in a bidirectional way. In this RTD approach, some of the tokens are replaced by plausible sample tokens from a generator network, and this approach facilitates models to learn from all input tokens. The model has a discriminator that facilitates distinguishing the replaced tokens.

9.6.1.5 Longformer

Longformer [13] is an MRC model for longer documents that utilizes some attention mechanisms, including sliding window attention and sparse global attention. Sliding window attention scales linearly with sequence length, so the models can handle longer documents with many sequences. The initialization of the tokenizer and model, input text tokenization, receiving the attention mask, retrieving the predictions, and converting them into the answer are the subprocedures in the Longformer model.

9.7 RESULTS AND DISCUSSION

The MRC span extraction task takes input as a question and a context. Here the context is taken as the abstract of this article and poses two questions to the model, one for the first execution (question = "What is MRC?" and question = "Neural Machine Reading Comprehension?"). The first question is the Natural form question, which conforms to the grammar of the natural language, and the second is the synthetic style question, which does not necessarily adhere to standard grammatical rules. First and foremost, *BertForQuestionAnswering* and *BertTokenizer* classes must be imported from the Transformers to access the pre-trained model. The SQUAD [14,15] model was trained using the benchmarked dataset.

from transformers import BertForQuestionAnswering

model = BertForQuestionAnswering.from_pretrained('bert-large-uncased-whole-word-masking-finetuned-squad')

from transformers import BertTokenizer

tokenizer = BertTokenizer.from_pretrained('bert-large-uncased-whole-word-masking-finetuned-squad')

The initial stage of the MRC is embedding; here, the BERT model has three embeddings, including input embedding, segment embedding, and position embedding. The input embedding converts the input context and question to a numerical format as vectors. The segment embedding facilitates the model's ability to distinguish between the question and context and the sequences in the context. Finally, the position embedding helps identify each token's position in the input. Then these embeddings pass to the encoder-decoder layers of the transformer, which then extract the most relevant features from the context and question. The model retrieves the context's start and end logits of the plausible answers. The answers for the first and second questions with starting tensors are *{'Answer_start': tensor(50), 'text' : 'Answer' : 'machine reading comprehension'}* and *{'Answer_start': tensor(89), 'text' : 'Answer' : 'neural machine reading comprehension has built reading comprehension models by employing the advancements of deep neural networks'}*, respectively.

9.8 CONCLUSION AND FUTURE ENHANCEMENT

Machine reading comprehension has the potential to provide accurate answers from Natural Language documents. The article incorporates the MRC architecture, modules, tasks, benchmarked datasets, and PEMs. A new classification of the MRC tasks was included with representative datasets. Furthermore, the deep neural network, Transformer, BERT, and BERT-based models are explored.

REFERENCES

1. Lehnert, W.G.; 1977; The *Process* of Question *Answering*; Ph.D. thesis; Yale University.
2. Hirschman, L.; Marc, L.; Eric, B.; and John, D. B.; 1999; Deep Read: A Reading Comprehension System; In *Proceedings of the 37th Annual Meeting of the Association for Computational Linguistics on Computational Linguistics*; pp 325–332.
3. Kenton, L.; Shimi, S.; Tom, K.; Ankur, P.; Dipanjan, D.; and Jonathan, B.; 2017; *Learning Recurrent Span Representations for Extractive Question Answering*; arXiv:1611.01436v2 [cs.CL]; pp. 1–9.
4. Haichao, Z.; Furu, W.; Bing, Q.; and Ting, L.; 2018; Hierarchical Attention Flow for Multiple-Choice Reading Comprehension; *The Thirty-Second AAAI Conference on Artificial Intelligence (AAAI-18)*; 32(1); pp. 6077–6084.
5. Zhipeng, C.; Yiming, C.; Wentao, M.; Shijin, W.; and Guoping, H.; 2018; *Convolutional Spatial Attention Model for Reading Comprehension with Multiple-Choice Questions*; arXiv:1811.08610v1 [cs.CL]; pp. 1–8.
6. Shuailiang, Z.; Hai, Z.; Yuwei, W.; Zhuosheng, Z.; Xi, Z.; and Xiang, Z.; 2019; *Dual Co-Matching Network for Multi-choice Reading Comprehension*; arxiv.org: 1901.09381v2 [cs.CL]; pp. 1–8.

7. Yiqing, Z.; Hai, Z.; and Zhuosheng, Z.; 2019; Examination-style Reading Comprehension with Neural augmented Retrieval; *International Conference on Asian Language Processing (IALP)*; IEEE; pp. 182–187.

8. Jacob, D.; Ming-Wei, C.; Kenton, L.; and Kristina, T.; 2019; *BERT: Pre-training of Deep Bidirectional Transformers for Language Understanding*; arXiv:1810.04805; pp. 1–16.

9. Victor, S.; Lysandre, D.; Julien, C.; and Thomas, W.; 2020; *DistilBERT: A Distilled Version of BERT: Smaller, Faster, Cheaper and Lighter*; arXiv:1910.01108v4 [cs.CL]; pp. 1–5.

10. Zhenzhong, L.; Mingda, C.; Sebastian, G.; Kevin, G.; Piyush, S.; and Radu, S.; 2020; *Albert: A Lite Bert for Self-Supervised Learning of Language Representations*; arXiv:1909.11942v6; pp. 1–17.

11. Yinhan, L.; Myle, O.; Naman, G.; Jingfei, D.; Mandar, J.; Danqi, C.; Omer, L.; Mike, L.; Luke, Z.; and Veselin, S.; 2019; *RoBERTa: A Robustly Optimized BERT Pre-training Approach*; arXiv:1907.11692v1; pp. 1–13.

12. Kevin, C.; Minh-Thang, L.; Quoc, V. L.; and Manning, C. D.; 2020; *ELECTRA: Pre-Training Text Encoders as Discriminators rather than Generators*; arXiv:2003.10555v1 [cs.CL]; pp. 1–18.

13. Iz, B.; Matthew, E. P.; and Arman, C.; 2020; *Longformer: The Long-Document Transformer*; arXiv:2004.05150v2 [cs.CL]; pp. 1–17.

14. Pranav, R.; Jian, Z.; Konstantin, L.; and Percy, L.; 2016; *SQuAD: 100,000+ Questions for Machine Comprehension of Text*; arXiv:1606.05250v3 [cs. CL]; pp. 1–10.

15. Pranav, R.; Robin, J., and Percy, L.; 2018; *Know What You Don't Know: Unanswerable Questions for SQuAD*; arXiv:1806.03822 [cs.CL].

16. Ashish, V.; Noam, S.; Niki, P.; Jakob, U.; Llion, J.; Aidan, N. G.; Łukasz, K.; and Illia, P.; 2017; Attention is All You Need. In *Advances in Neural Information Processing Systems*, pp. 5998–6008.

17. Hermann, K. M.; Toma's, K.; Edward, G.; Lasse, E.; Will, K.; Mustafa, S.; and Phil, B.; 2015; *Teaching Machines to Read and Comprehend*; *Advances in Neural Information Processing Systems*; pp. 1693–1701.

18. Petr, B.; 2015; Modeling of the Question-Answering Task in the Yodaqa System; *International Conference of the Cross-Language Evaluation Forum for European Languages*; Springer; pp. 222–228.

19. Yi, Y.; Wen-tau, Y.; and Christopher, M.; 2015; Wikiqa: A Challenge Dataset for Open-Domain Question Answering; *Proceedings of the 2015 Conference on Empirical Methods in Natural Language Processing*; pp. 2013–2018.

20. Alexander, H. M.; Adam, F.; Jesse, D.; Antoine, B.; and Jason, W.; 2016, *KeyValue Memory Networks for Directly Reading Documents*; arXiv:1606.03126v2 [cs.CL]; pp. 1–10.

21. Takeshi, O.; Hai, W.; Mohit, B.; Kevin, G.; and David, M.; 2016; Who did What: A Large-Scale Person-Centered Cloze Dataset; *Proceedings of the 2016 Conference on Empirical Methods in Natural Language Processing; Association for Computational Linguistics*, pp. 2230–2235.

22. Payal, B.; Daniel, C.; Nick, C.; Li, D.; Jianfeng, G.; Xiaodong, L.; Rangan, M.; Andrew, M.; Bhaskar, M.; Nguyen, T.; Rosenberg, M.; Xia, S.; Alina, S.; Saurabh, T.; and Tong, W.; 2016; *MS MARCO: A Human Generated Machine Reading COmprehension Dataset*; CoRR 2016; abs/1611.09268; pp. 1–11.

23. Adam, T.; Wang, T.; Yuan, X, Harris, J., Sordoni, A.; Bachman, P.; and Suleman, K., 2017, NewsQA: A Machine Comprehension Dataset; *Proceedings of the 2nd Workshop on Representation Learning for NLP*; Association for Computational Linguistics, pp. 191–200.

24. Hill, F.; Antoine, B.; Sumit, C.; and Jason, W.; 2016; The Goldilocks Principle: Reading Children's Books with Explicit Memory Representations; *4th International Conference on Learning Representations*, ICLR; https://arxiv.org/pdf/1511.02301.pdf; pp. 1–13.

25. Ondrej, B.; Rudolf, K.; and Jan, K.; 2016; *Embracing Data Abundance: BookTest Dataset for Reading Comprehension*; arXiv:1610.00956v1 [cs.CL]; pp. 1–10.

26. Radu, S.; and Nan, D.; 2016.; *Building Large Machine Reading-Comprehension Datasets using Paragraph Vectors*; arXiv preprint arXiv:1612.04342; pp. 1–10.

27. Tapaswi, M.; Yukun, Z.; Rainer, S.; Antonio, T.; Raquel, U.; and Sanja, F.; 2016; Movieqa: Understanding Stories in Movies Through Question-Answering; *Proceedings of the IEEE Conference on Computer Vision and Pattern Recognition*; pp. 4631–4640.

28. Mandar, J.; Eunsol, C.; Daniel, W.; and Luke, Z.; 2017; *TriviaQA*: A Large Scale Distantly Supervised Challenge Dataset for Reading Comprehension; *Proceedings of the 55th Annual Meeting of the Association for Computational Linguistics*; pp. 1601–1611.

29. Lai, G.; Qizhe, X.; Hanxiao, L.; Yiming, Y.; and Eduard, H.; 2017; RACE: Large-Scale Reading Comprehension Dataset from Examinations; *Proceedings of the 2017 Conference on Empirical Methods in Natural Language Processing*; pp. 785–794.

30. Bhuwan, D.; Kathryn, M.; and William, W. C.; 2017; *QUASAR: Datasets for Question Answering by Search and Reading*; arXiv:1707.03904v2 [cs.CL]; pp. 1–10.

31. Matt, D.; Levent, S.; Mike, H.; Volkan, C.; and Kyunghyun, C.; 2017; *SearchQA: A New Q&A Dataset Augmented with Context from a Search Engine*; arXiv:1704.05179v3 [cs.CL]; pp. 1–5.

32. Tomáš, K.; Jonathan, S.; Phil, B.; Chris, D.; Hermann, K. M.; Gábor, M.; and Edward, G.; 2018; *The NarrativeQA Reading Comprehension Challenge.* Transactions of the Association for Computational Linguistics; MIT Press: Cambridge, MA; pp. 317–328.

33. Johannes, W.; Stenetorp, P.; and Sebastian, R.; 2018; *Constructing Datasets for Multi-Hop Reading Comprehension Across Documents*; *Transactions of the Association for Computational Linguistics*; pp. 287–302.

34. Aniruddha, K.; Minjoon, S.; Dustin, S.; Jonghyun, C.; Ali, F.; and Hannaneh, H.; 2017; Are You Smarter Than a Sixth-Grader? Textbook Question Answering for Multimodal Machine Comprehension; *Proceedings of the IEEE Conference on Computer Vision and Pattern Recognition*, pp. 4999–5007.

35. Mohit, I.; Varun, M.; Anupam, G.; Yogarshi, V.; Jordan, B. G.; Hal, D.; and Larry, D.; The Amazing Mysteries of the Gutter: Drawing Inferences Between Panels in Comic Book Narratives; *Proceedings of the IEEE Conference on Computer Vision and Pattern Recognition*; pp. 7186–7195.

36. Choi, E.; He, H..; Mohit, I.; Mark, Y.; Wen-tau, Y.; Yejin, C.; Percy, L.; and Luke, Z.; 2018; QuAC: Question Answering in Context; *Proceedings of the 2018 Conference on Empirical Methods in Natural Language Processing; Association for Computational Linguistics*; pp. 2174–2184.

37. Siva, R.; Danqi, C.; Manning; and C., D.; 2019; *CoQA: A Conversational Question Answering Challenge*; Transactions of the Association for Computational Linguistics; MIT Press: Cambridge, MA; pp. 249–266.

38. Zhilin, Y.; Peng, Q.; Saizheng, Z.; Yoshua, B.; William, W., Cohen; Ruslan, S.; and Manning, C. D.; 2018; *HOTPOTQA*: A Dataset for Diverse, Explainable Multi-hop Question Answering; *Proceedings of the 2018 Conference on Empirical Methods in Natural Language Processing*; Association for Computational Linguistics, pp. 2369–2380.

39. Amrita, S.; Rahul, A.; Mitesh, M. K.; and Karthik, S.; 2019, *DuoRC: Towards Complex Language Understanding with Paraphrased Reading Comprehension*; arXiv:1804.07927v4 [cs.CL]; pp. 1–16.

40. Qizhe, X..; Guokun, L..; Zihang, D.; and Eduard, H.; 2018; Large-scale Cloze Test Dataset Created by Teacher; *Proceedings of the 2018 Conference on Empirical Methods in Natural Language Processing*; Association for Computational Linguistics: Brussels, Belgium; pp. 2344–2356.

41. Sheng, Z.; Xiaodong, L.; Jingjing, L.; Jianfeng, G.; Kevin, D.; and Van, D. B.; 2018; *Record: Bridging the Gap Between Human and Machine Commonsense Reading Comprehension*; arXiv preprint arXiv:1810.12885; pp. 1–14.

42. Simon, S.; and Walter, D.; 2018; *CliCR: A Dataset of Clinical Case Reports for Machine Reading Comprehension*; arXiv:1803.09720; pp. 1551–1563.

43. Quentin, G.; and Julien, P.; 2018; *ReviewQA: A Relational Aspect-Based Opinion Reading Dataset*; arXiv preprint arXiv:1810.12196; pp. 1–10.

44. Isaac, C.; Oren, E.; Tushar, K.; Ashish, S.; Carissa, S.; and Oyvind, T.; 2018; *Think you have Solved Question Answering? Try ARC, the AI2 Reasoning Challenge.* arXiv 2018, abs/1803.05457; pp. 1–10.

45. Todor, M.; and Anette, F.; 2018; Knowledgeable Reader: Enhancing Cloze-Style Reading Comprehension with External Commonsense Knowledge; *Proceedings of the 56th Annual Meeting of the Association for Computational Linguistics*; pp. 821–832.

46. Tushar, K.; Ashish, S.; and Peter, C.; 2018; Scitail: A Textual Entailment Dataset from Science Question Answering; *Thirty-Second AAAI Conference on Artificial Intelligence*; pp. 1–9.

47. Daniel, K..; Snigdha, C.; Michael, R.; Shyam, U.; and Dan, R.; 2018; Looking Beyond the Surface: A Challenge Set for Reading Comprehension over Multiple Sentences; *Proceedings of the 2018 Conference of the North American Chapter of the Association for Computational Linguistics: Human Language Technologies*; pp. 252–262.

48. Semih, Y.; Aykut, E.; Erkut, E.; and Cinbi, N. I.; 2018; RecipeQA: A Challenge Dataset for Multimodal Comprehension of Cooking Recipes; *Proceedings of the 2018 Conference on Empirical Methods in Natural Language Processing*; pp. 1358–1368.

49. Park, D.; Yonghwa, C.; Daehan, K.; Minhwan, Y.; Seongsoon, K.; and Jaewoo, K.; 2019; Can machines learn to comprehend scientific literature? *IEEE Access*, 7, 16246–16256.

50. Simon, O.; Ashutosh, M.; Michael, R.; Stefan, T.; and Manfred, P.; 2018; MCScript: A Novel Dataset for Assessing Machine Comprehension Using Script Knowledge; *Proceedings of the Eleventh International Conference on Language Resources and Evaluation (LREC 2018)*; pp. 3567–3574.

51. Bhavana, D.; Lifu, H.; Niket, T.; Wen-tau, Y.; and Peter, C.; 2018; Tracking State Changes in Procedural Text: A Challenge Dataset and Models for Process Paragraph Comprehension; *Proceedings of the 2018 Conference of the North American Chapter of the Association for Computational Linguistics*: Human Language Technologies, pp. 1595–1604.

52. Tom, K.; Jennimaria, P.; Olivia, R.; Michael, C.; Ankur, P.; Chris, l.; Danielle, E.; Illia, P.; Jacob, D.; Kenton, L.; Kristina, T.; Llion, O.; Matthew, K.; Ming-Wei, C.; Andrew, M. D.; Jakob, U.; Quoc, L.; and Slav, P.; 2019; Natural questions: A benchmark for question answering research; *Transactions of the Association for Computational Linguistics*; 7(1); pp. 453–466.

53. Dian, Y.; Jianshu, C.; Dong, Y.; Yejin, C.; and Claire, C.; 2019; *DREAM: A Challenge Dataset and Models for Dialogue-Based Reading Comprehension*; arXiv:1902.00164v1 [cs.CL]; pp. 217–231.

54. Marzieh, S.; Max, B.; Patrick, L.; Sameer, S.; Tim, R.; Mike, S.; Guillaume, B.; and Sebastian, R.; 2018; Interpretation of Natural Language Rules in Conversational Machine Reading; *Proceedings of the 2018 Conference on Empirical Methods in Natural Language Processing*; Association for Computational Linguistics: Brussels, Belgium, pp. 2087–2097.

55. Alon, T.; Jonathan, H.; Nicholas, L.; and Jonathan, B.; 2019, CommonsenseQA: A Question Answering Challenge Targeting Commonsense Knowledge; *Proceedings of the 2019 Conference of the North American Chapter of the Association for Computational Linguistics: Human Language Technologies*; pp. 4149–4158.

56. Dheeru, D.; Yizhong, W.; Pradeep, D.; Gabriel, S.; Sameer, S.; and Matt, G.; 2019; DROP: A Reading Comprehension Benchmark Requiring Discrete Reasoning Over Paragraphs. *Proceedings of NAACL*; pp. 2368–2378.

57. Jian, L.; Leyang, C.; Hanmeng, L.; Dandan, H.; Yile, W.; and Yue, Z.; 2020; *LogiQA: A Challenge Dataset for Machine Reading Comprehension with Logical Reasoning*; arXiv:2007.08124v1 [cs.CL]; pp. 1–7.

58. Timo, M.; Anthony, R.; Raghavan, J.; and Malte, P.; 2020; *COVID-QA: A Question Answering Dataset for COVID-19*; The White House Office of Science and Technology Policy; Call to Action to the Tech Community on New Machine Readable COVID-19 Dataset.

59. Weihao, Y.; Zihang, J.; Yanfei, D.; and Jiashi, F.; 2020; *Reclor: A Reading Comprehension Dataset Requiring Logical Reasoning*; *ICLR*; arXiv:2002.04326; pp. 1–26.

Chapter 10

Online subjective question-answering system necessity of education system

Madhav A. Kankhar, Bharat A. Shelke, and C. Namrata Mahender
Dr. Babasaheb Ambedkar Marathwada University

10.1 INTRODUCTION

Academic performance is measured based on examinations; the examinations can be subjective or objective; and in the COVID-19 pandemic, one works in an online or virtual model. Presently, manual checking of subjective answers is a tedious approach. Most of the systems are present, which easily evaluate objective-type questions. Techniques developed in machines to give predetermined correct answers. It is only helpful for objective types of examinations. The backbone of many universities and state boards is subjective examination. Students attend the subjective exam, and depending upon the answer, knowledge, and academic performance, the moderator will access the students' performance and grade them. Manual evaluation for subjective examinations is a very tedious and time-consuming process. The evaluation of answers varies from moderator to moderator, and sometimes relations among students and faculty also affect the evaluation process. There are many more reasons, such as mood, bulk checking, etc. The result is affected by it; the main aim of this work is to automate the subjective examination system using the natural language process (Girkar, 2021).

10.1.1 Brief on NLP (natural language processing)

Natural language processing (NLP) is commonly used for communication, and it can be spoken, signed, or written. Natural language differs from constructed and formal languages, such as computer programming languages or the "Languages" used to study formal logic, particularly mathematical logic.

NLP is a procedure that allows machines to take on more human characteristics, thereby shortening the gap between humans and machines (Horowitz, 2012). Natural language understanding is sometimes referred to as an AI-complete problem because it appears to necessitate extensive knowledge of the outside world and manipulative abilities.

DOI: 10.1201/9781003244332-10

NLP is a computerized text evaluation technique based on two factors. One is a set of theories, and the other is a set of technologies. Because there is so much research and development, it is difficult to define a single definition that will satisfy everyone. However, some characteristics can be useful in describing NLP.

10.1.1.1 Stages of NLP

In an NLP system that accepts spoken input, sound waves are analyzed and encoded into a digitized signal for interpretation by various rules compared to the language model (refer to Figure 10.1). The below stages are concerned with written language processing, as they are familiar with verbal, except in verbal it starts with phonetics.

10.1.1.1.1 Morphology and lexical analysis

A language's lexicon is its vocabulary, including its words and expressions. Morphology depicts the analysis, identification, and description of word structure. Words are widely regarded as the minor units of syntax. The rules and principles that govern the sentence structure of any

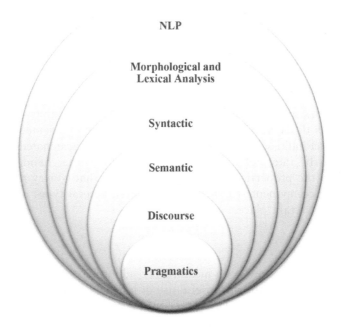

Figure 10.1 Stages of NLP.

individual language are referred to as syntax. The lexical analysis aims to divide the text into paragraphs, sentences, and words. The lexical analysis cannot be carried out independently of the morphological and syntactic analyses.

10.1.1.1.2 Syntactic

This level focuses on analyzing the words in a sentence to determine the sentence's grammatical structure. This necessitates the use of both grammar and a parser. This level of processing produces a sentence representation that reveals the structural dependency relationships between the words. Multiple grammars can be used, influencing the choice of a parser. Because not all NLP applications require an entire sentence parse, the remaining challenges in parsing prepositional phrase attachment and conjunction scoping no longer confuse applications that rely on phrasal and clausal dependencies.

10.1.1.1.3 Semantic

Semantic processing determines a sentence's possible meaning by focusing on the interactions between word-level meanings in the sentence. This level of processing can include semantic disambiguation of words with multiple senses, similar to how syntactic disambiguation of words with multiple parts of speech (POH) is accomplished at the syntactic level. Semantic disambiguation allows only one sense of a polysemous word to be chosen and included in the sentence's semantic representation.

10.1.1.1.4 Discourse

While syntax and semantics deal with sentence length units. NLP's discourse-level deals with text units that are longer than a sentence. It does not interpret multi-sentence texts as a collection of concatenated sentences, which can be interpreted independently. On the other hand, discourse focuses on the properties of the text as a whole that convey meaning by connecting component sentences. At this level, several types of discourse processing can take place.

1.1.1.5 Pragmatic

This level explains how extra meaning is read into texts without encoding. It is concerned with the purposeful use of language in situations and uses its context in addition to the contents of the text for understanding. This necessitates extensive world knowledge and an understanding of intentions, plans, and goals.

10.1.2 Question

The answer to a question is never found in a dictionary, but the response requires a dictionary, and sometimes the entire universe must be searched for just a response. A question is a noun sentence written or spoken to elicit information (Gite et al., 2013). By definition, a question is a linguistic expression used to request information, or the request for information, but understanding this request in written or verbal form is a complex task; for example, if we consider the written language, when a question mark is seen, that sentence is considered as an interrogative sentence, but many times it is misleading as the rhetorical question does not require an answer but is often terminated by a question mark. In contrast, statements asking for information may not be phrased as a question.

E.g., why are you so stupid?

Are you kidding me?

10.1.3 Answer

The information is requested as a question, and the answer is a response to the question. According to the definition, the answer is a noun, a thing that is said, written, or done to deal with it, or as a reaction, statement, or situation. An answer to a specific question may employ more than one mode of expression. As a result, the responses are sent to the specific aspect relevant to each user (Khillare et al., 2014).

While there are many ways to express the correct answer to a question, not all of them will meet the needs of the questioners.

Example: Who was Dr. Rajendra Prasad?

a. First President of India.
b. Freedom Fighter.
c. Freedom fighter and first president of India.

10.1.4 Question-answering system

The user submits various queries to the question-answering system to receive accurate responses from the system. QA is an important solution for obtaining valid and accurate answers to user questions in natural language rather than a query (Khillare et al., 2014). A question-answer system is a type of information retrieval system. A question-answering system attempts to find the correct answer to a natural language question given a set of documents.

Question-answering is a multidisciplinary field encompassing information technology, artificial intelligence, NLP, knowledge and database management, and cognitive science.

For designing a question, skill is required, and in the education system, many taxonomies are there but widely and majorly in fields (science, arts, engineering, etc.). Bloom's taxonomy is used. A brief description of Bloom's taxonomy and Lenhert's taxonomy is discussed here.

10.1.4.1 Bloom's taxonomy

Bloom's taxonomy was created in 1965 by Dr. Benjamin Bloom to promote higher levels of thinking in the educational domain (Das, 2008). The goal was to encourage higher-order thinking skills in students, such as analyzing and evaluating, rather than simply recalling facts. Bloom's taxonomy has six levels, which are listed in order of lowest to highest. These levels are discussed in the following sections.

10.1.4.1.1 Knowledge

Questions assess a student's ability to memorize and recall terms, facts, and details without necessarily understanding the concept: remembering and recalling relevant, previously learned information is here defined as knowledge.

The main keyword of this level is Memorize, Define Identity Repeat, Recall, State, Write, List, and Name.

10.1.4.1.2 Comprehension

These questions assess the student's ability to summarize and describe information in their own words without necessarily relating it to anything else. This level's main keywords are Describe, Distinguish, Explain, Interpret, Predict, Recognize, and Summarize.

10.1.4.1.3 Application

The application of previously learned information to new and concrete situations to solve problems with single or best answers. This level's main keywords are Apply, Compare, Contrast, Demonstrate, Examine, Relate, Solve, and use.

10.1.4.1.4 Analysis

These questions encourage students to divide the material, describe patterns and relationships between parts, subdivide information, and demonstrate how it is put together. Analyze, Differentiate, Distinguish, Explain, Infer, Relate, Research, and Separate are keywords.

10.1.4.1.5 Synthesis

These questions challenge students to create something new by combining ideas from various sources to form a new whole. The keywords are arranged: Combine, Create, Design, Develop, Formulate, Integrate, and Organize.

10.1.4.1.6 Evaluation

Questions encourage students to form opinions and value judgments about issues based on specific criteria. Assess, Critique, Determine, Evaluate, Judge, Justify, Measure, and Recommend are keywords.

All of the preceding stages are part of the cognitive domain.

10.1.4.2 Lenhert taxonomy

The automatic question-answering system based on story understanding grew. The automation of question-answering was first attempted by Lenhert's in 1978. The name of the system was QUALM. QUALM was developed to replicate the answer to the question as a human would. The questions were based on short stories. The standard classification of questions in English is the five Ws. It is a standard way to represent questions that we learned in school. The essential elements of Ws are what, who, whom, when, why, where, and how. H represents perversity in English spelling. The five Ws are the best way to think about questions. Robinson and Rackstraw dedicate the volumes to analyzing Wh-words. (Pomerantz, 2005) The questions are based on Wh-words and also answer the question. Robinson and Rackstraw mentioned that Wh-words are a set of lexically marked interrogative words. Robinson and Rackstraw Lexical set to mean a discrete set of words used in a similar linguistic environment. Wh-words are used to make interrogative sentences. Any sentences consisting of a Wh-word are considered questions.

Robinson and Rackstraw add one class to this group and differentiate between open and closed questions by phrasing the words what and which. The word indicates an indefinite set, recognizing a specific object from a finite set of objects. According to Robinson and Rackstraw's taxonomy of Wh-questions, they are as follows:

Who
What
When
Where
Why
Which
How

Another problem related to the question domain is the number of sentences containing the Wh-word.

1. Questions are not always framed using Wh-words.
2. Sentences are framed using Wh-words that are necessarily questions.

It means that questions are framed as statements but accepted as questions.

10.1.5 Types of question-answering

The question-answering system is based on multiple domains discussed in the following sections.

10.1.5.1 Based on domain

I. **Open Domain**

Open domains work with web-based modules; the web is one of the best sources of information; open-domain question-answering systems use search engines such as Google, Yahoo, Ask.com, Rediffmail, and others to provide information. Open-domain QA systems are web-based systems that have access to a lot of data. The first web-based question-answering system, dubbed "START," was created in 1990 (START, 2015).

II. **Closed Domain**

Closed domains predate open domains. The benefit of this system is that it deals with precise data that does not usually contain ambiguous terms and can thus be processed more efficiently. For example, on the BCCI blog (Board of Control for Cricket in India), all questions and answers are about cricket. When modeling a specific domain, the relevant concepts and their relationships are predetermined and explicitly stored, reducing the number of possible interpretations of a question. Rather than learning a complex query language such as SQL, Users can query the data using natural language, and the system automatically converts the user's question into a query (Dornescu, 2012).

10.1.5.2 Based on question types

I. **Factoid Question**

These are simple, fact-based questions that require answers in a single short phrase or sentence, such as "Who is the producer of the movie XYZ?" The factoid-type questions usually begin with the letter wh. Current QASs have demonstrated adequate performance in answering factoid-type questions. Most factoid-type questions have expected answer types that are generally named entities that can be traced in documents using named entity tagging software. The type of question determines them. As a result, good accuracy is possible (Kolomiyets et al., 2011).

II. List Types Question

In order to answer the list questions, you must provide a list of entities or facts, such as listing the names of employees earning more than 50,000 rupees per year; QASs consider such questions to be a series of factoid questions ten times in a row. QAS disregards previous answers when posing new questions. In general, QASs report difficulty determining the threshold value for the number or quantity of the entity asked in list-type questions; in list-type questions, QASs observe a problem in determining the threshold value for the number or quantity of the entity asked (Indurkhya & Damerau, 2010).

III. Definition Types Question

Definition questions of the form 'What is X' are frequently asked on the Internet. This type of question is commonly used to obtain information about an organization or thing. In general, dictionaries and encyclopedias are the best resources for this type of information. However, due to non-instantaneous updates, these resources frequently do not contain the most recent information about a specific organization or do not yet contain a definition of a new organization. As a result, the user has developed the habit of searching the Web for definitions (Trigui et al., 2010).

IV. Hypothetical Type Question

Hypothetical questions seek information about a hypothetical event. They typically begin with, "What would happen if QASs required knowledge retrieval techniques to generate answers?" Furthermore, the answers to these questions are subjective. These questions have no specific answers (Kolomiyets et al., 2011).

V. Casual Question

Causal questions necessitate explanations for an entity. As in the case of factoid-type questions, the answers are not named entities. QAS requires advanced NLP techniques to analyze text at the pragmatic and discourse levels to generate answers. Such questions are asked by users who want explanations, reasons, elaborations, and responses to specific events or objects (Higashinaka & Isozaki, 2008).

VI. Confirmation Question

Answers to confirmation questions must be in the form of a yes or no. Systems require an inference mechanism, world knowledge, and common sense reasoning (Mishra & Jain, 2016).

10.1.5.3 Based on data source

I. Information retrieval-based

Text-based or IR-based QA is based on the massive amount of information available on the web as text or ontologies. By retrieving text snippets from the web or corpora, this QA responds to the user's query. A sample factoid query is "Who is the president of India?" The answer is "President Ramnath Kovind is on vacation in London." IR methods extract passages from these documents directly, guided by the question. The QA framework structure explains the IR QA system as it performs its function (Jurafsky & Martin, 2015).

II. Knowledge-based

Giving answers to natural language questions by mapping them to a query over an ontology is what knowledge-based question-answering entails. Whatever logical form is derived from the mapping to retrieve facts from databases, any complex structure, such as scientific facts or geospatial readings, can be used as the data source, requiring complex logical or SQL queries (Jurafsky & Martin, 2015).

III. Hybridized or multiple knowledge sources

This entails utilizing structured knowledge bases and text datasets to provide answers to queries. DeepQA is a practical hybridized QA system that is information source-based. This system extracts a wide range of senses from the query, including entities, parses, relations, and ontological information, and then retrieves candidate answers from textual sources and knowledge bases (Chu-Carroll et al., 2003).

10.1.5.4 Based on forms of answer generation

I. Extracted answers

Extracted answers are classified into three types: answers in the form of sentences, paragraphs, and multimedia. Answers in sentence form are based on documents that have been retrieved and divided into individual sentences. The user's query heavily influences the answer form. The user is shown the sentence with the highest ranking. Factoids or confirmation questions typically have extracted answers (Mishra & Jain, 2016).

II. Generated answers

Generated answers are classified into conformational, yes or no, opinionated, or dialogue answers. Confirmation questions have yielded yes or no answers through confirmation and reasoning. Opinionated answers are generated by QA systems that assign star ratings to objects or features of objects. The Dialog QA system returns answers to questions (Mishra & Jain, 2016).

10.1.5.5 Based on language

I. Monolingual

The user's query, resource documents, and system response are all expressed in a single language in this system. START, Freebase, and others process user queries expressed in English, using facts from built-in databases articulated in English, and return answers in the language form in which the questions were asked (Fahmi, 2009).

II. **Cross-lingual**

The user's query and resource documents are expressed in different languages in this system, and the query is interpreted into the language of the resource documents before the search. Google Knowledge Graph converts queries entered in other languages into English and returns an answer in English. Quantum is an English-French cross-lingual system that translates users' queries from French to English, processes them in English, and translates the results back to French (Plamondon & Foster, 2003).

III. **Multilingual**

This system expresses the user's query and resource documents in the same language. The query search is performed in the same language as the query; e.g., the system developed by The Health on the Net Foundation supports English, French, and Italian, while the DIOGENE system supports only two languages: English and Italian, but unlike cross-lingual systems, the question is asked in the same language as the query (Magnini et al., 2003).

10.1.5.6 Based on application domain

I. **General domain QASs**

QAS in the public domain question-answering systems answer domain-independent questions. In general, QASs search for answers within an extensive document collection. A large repository of questions can be asked in public domain QASs. Question-answering systems use general ontology and world knowledge (Lam et al., 2006).

II. **Restricted domain QASs**

Restricted domain QASs answer domain-specific questions (Molla and Vicedo, 2007). Answers are found by searching domain-specific document collections. Because the repository of question patterns is small, the systems can accurately answer questions. QASs use domain-specific ontologies and terminology. The quality of the responses is expected to be higher. Various restricted domain QASs have been developed in the literature, such as temporal domain QAS, geospatial domain QAS, medical domain QAS, patent QAS, community-based QAS, etc. Various restricted domain QASs can be combined to form General domain QASs (Lopez et al., 2011).

10.1.6 Approaches used for developing QA system

10.1.6.1 Linguistic approach

The linguistic approach encompasses natural language text, linguistic knowledge, and shared knowledge. Tokenization, POS tagging, and parsing are examples of linguistic techniques. These were applied to a user's query to turn it into a precise query that simply extracts the appropriate response from the structured database (Sasikumar, & Sindhu, 2014).

10.1.6.2 Statistical approach

The availability of massive amounts of data on the Internet has increased the importance of statistical approaches. A statistical learning method outperforms other approaches in terms of results. Online text repositories and statistical approaches are not limited to structured query languages and can be used to formulate queries in natural language. Statistical techniques such as support vector machine classifiers, Bayesian, and maximum entropy models are commonly used in QA systems (Dwivedi & Singh, 2013).

10.1.6.3 Pattern matching

The pattern-matching approach deals with the expressive power of text patterns and replaces the complex processing involved in other computing approaches. Most pattern-matching QA systems use the surface text pattern, with some relying on response generator templates (Dwivedi & Singh, 2013).

10.1.7 Components of question-answering

The QA system has standard architecture consisting of three components linked sequentially: Question Processing which identifies the type of put question. Passage Retrieval, which extracts and ranks exact responses from previously retrieved passages and Answer Extraction, which extracts and ranks a small number of relevant passages from the underlying speech transcriptions (Marvin, 2013).

Auto Question Generation (AQG): In most question-answering systems, a question bank POS tagger is used to help in question generation from the given sentences. It requires in-depth knowledge of tags for proper rule development when designing a question.

Question Processing (QP): Given a natural language question as input, the question processing module's overall function is to analyze and process the question by creating some representation of the information requested. As a result, the question processing module must (Allam & Haggag, 2012):

Analyze the question to represent the primary information needed to answer the user's question.

Classify the question type, typically using a taxonomy of possible questions already coded into the system, which leads to the expected answer type via some shallow semantic processing of the question.

Reformulate the question to improve its phrasing and transform it into queries for the information retrieval search engine.

Passage Retrieval (PR): The PR process is divided into two major steps: (a) in the first step, all non-stop question words are sorted in descending order of priority; and (b) in the second step, the set of keywords used for retrieval and their proximity are dynamically adjusted until the number of retrieved passages is sufficient (Marvin, 2013).

Answer Extraction (AE): The AE component identifies and extracts the answer from the relevant passage set most likely to respond to the user's question. The parser recognizes the answer candidates within the paragraphs. As a result, once an answer candidate has been identified, a set of heuristics is used to extract only the relevant word or phrase that answers the question (Allam & Haggag, 2012).

Model Answer Generation (MG): The Faculty or the administrator develops model answers for the question generated, which act as guidelines for evaluating the answers of candidates' students.

Answer Evaluation (AE): All the answers from the candidates are matched with the model answers based on the similarity in the answers evaluated. There are several ways to boost confidence in the correctness of an answer. One method used a lexical resource such as WordNet to validate that a candidate's response was of the correct answer type. Furthermore, specific knowledge sources can be used as a second opinion to validate answers to questions within specific domains (Allam & Haggag, 2012).

System Measurement (SM): F-score, recall, accuracy, and precision are a few system performance measurements mainly used for question-answering systems.

10.1.8 Need for question-answering system

Essentially, the answering question system will provide a platform for users' needs to accurately and effectively answer the posed question. The second important need that has been identified as a result of globalization's impact is education, which is now available to users at any time, from any location, and for any type of cause from various reputable institutes. This benefits learners and trainers by allowing them to obtain a course of their choice in essay mode. However, from the standpoint of education, the learning process is incomplete without an exam, as passing a course requires passing. When we look at the exam scenario of any large institution or named educational institute in any part of the world, subjective examination plays a vital role in skill development. It also increases the tendency toward problem-solving in a better way and aids in enhancing the linking capacity for better decision-making.

10.2 QUESTION TYPES FOLLOW INTO TWO CATEGORIES

10.2.1 Objective examination

The term objective test refers to tests whose scoring is based on personal judgments or opinions. Multiple-choice items (MCI), True/False items, matching items, transformation sentences, re-arrangement items, and fill in blanks or gap-filling are used in objective tests. Only a few approaches in the literature show an interest in open-cloze question generation.

The authors (Pino & Eskenazi, 2009) provided a hint in the form of an open-ended question. They discovered that the first few letters of a missing word hinted at the missing word. Their goal was to change the number of letters in a hint to change the difficulty level of questions, allowing students to learn vocabulary more quickly. (Agarwal, 2012) created an automated method for generating open-ended questions. Their method consisted of two steps: selecting relevant and informative sentences and identifying keywords in those sentences. His proposed system took cricket news articles as input and produced open-ended factual questions as output. Das and Majumder (2017) described a system for generating open-ended questions to assess learners' factual knowledge. They calculated the evaluation score using a formula based on the number of hints used by the students to provide correct answers. The multiword answer to the open-close question improves the system's appeal. Coniam (1997) proposed one of the oldest close test item generation techniques. He used word frequencies to analyze the corpus at various stages of development, including obtaining test item keys, generating test item alternatives, building close test items, and identifying good and bad test items. To create test items, he matched each test item's key's word frequency and POH with a similar word class and word frequency. Chen et al., (2006) created a semi-automated method for generating grammatical test items using NLP techniques. Their method entailed handcrafting patterns to find authentic sentences and distractors on the Internet, transforming them into grammar-based test items. The method produced 77% meaningful questions, according to their experimental results.

10.2.2 Subjective examination

Subjective or essay questions that allow the student to organize and present an original response: essay writing, composition writing, letter writing, reading aloud, completion type, and answer are examples of objective tests. Subjective evaluations are ideal for writing, reading, art history, philosophy, political science, and literature. Rozali et al. (2010) provided a survey of dynamic question generation and qualitative evaluation techniques and a description of related methods found in the literature. Deena et al. (2020) proposed a question generation method based on NLP and Bloom's taxonomy that generated subjective questions dynamically while reducing memory occupation. The main challenge of subjective assessment is proper scoring.

As a result, automatic subjective-answer evaluation is a current research trend in the educational system (Burrows et al., 2015). It reduces the time and effort required for assessment in the educational system. To test the correct option, objective-type answer evaluation is simple and requires a binary mode of assessment: true/false. However, subjective-answer evaluation does not produce adequate results due to its complexity. The following section discusses some related works on subjective-answer evaluation and grading techniques (Bin et al., 2008). The text categorization model and the K-nearest neighbor (KNN) classifier were used for automated essay scoring. Each essay was expressed using the Vector Space Model. They used words, phrases, and arguments as essay features, and the term frequency-inverse document frequency (TF-IDF) weight represented each vector. The cosine similarity method calculated essay scores and achieved 76 percent average accuracy using feature selection methods such as term frequency, TF-IDF, and information gain. Kakkonen et al., (2008) proposed an automatic essay grading system that compares learning materials to teacher-graded essays using three methods: Latent Semantic Analysis (LSA), Probabilistic (PLSA), and Latent Dirichlet Allocation (LDA). Their system outperformed the K-NN-based grading system.

10.2.3 Subjective examination-related work

Subjective examination literature survey on different languages is briefed in Table 10.1.

10.2.4 Why is subjective examination important

Educationists say that there cannot be a single indicator of knowledge. It depends on the parameters that one wants to gauge through the assessment. The following points are mostly considered when having a preference for subjective examination:

Subjective questions are useful in evaluating the depth of conceptual knowledge and the ability to articulate. The ability to construct arguments and justify positions on a particular topic or concept can also be tested effectively. With a better comprehension of concepts and the ability to critically analyze divergent viewpoints, subjective assessment is a better option. In subjective questions, students have to organize the answer using their own words and thoughts, enhancing their communication skill or command of communication is better.

10.2.5 Online education system

Online education has become an integral part of today, primarily due to the COVID-19 scenario. Teaching, learning, and evaluating are the main components of the online education system. In examinations, online education also plays a very vital role. Considering the online option of exams, mostly objective (MCQ)-based exams have a high priority compared to subjective form of the exam due to many limitations during the conduct of online subjective exams. Globally, the demand for online examinations has increased;

Table 10.1 Subjective examination literature survey on different languages

Sr.no.	Author and year	Techniques	Languages	Lacuna
1	Gupta et al., 2019	Deep neural network technique	Hindi, English	There is less work done on descriptive questions in multilingual language.
2	Seena et al., 2016	Sentence tokenization, lemmatization, pattern matching	Malayalam	To improve the system's efficiency, it is needed to work on anaphora resolution for the Malayalam question-answering system.
3	Dhiraj et al., 2015	Augmented reality, information extraction	Hindi, Marathi	Working on different tracking functionality is important to improve virtual content in a natural environment.
4	Vignesh and Sowmya, 2013	Rule-based approach	Tamil	The result may be improved by working on generating the grammatically correct question.
5	Carrino et al., 2019	TAR method, Squad Dataset	Spanish	It is needed to test the Squad dataset result with other languages also.
6	Thaker and Goel, 2015	Domain Selection and Ontology Construction Parse User Query Find information from constructed ontology based on parsed query	Urdu	It is important to work on auto-generated ontologies from a user query.
7	Kanaan et al., 2009	keyword matching	Arabic	There is a need to develop a cross-language question-answering system.
8	Mayeesha et.al., 2021	BERT	Bengali	There is a need to test results with another advanced language model other than the Bengali language to enhance the model.

we find all significant players are heading toward long answer options like Google quiz's Microsoft team. The main hurdle is auto evaluation of short and long answers.

Examples:

1.

11. _____means negative external trends or changes that can hinder a firms performance.

0   / 1 pt

Auto-graded

Threats	X

Correct answers: threat

2.

13. Sometimes plan prepared become _____even before there implementation.

0   / 1 pt

Auto-graded

Obselete	X

Correct answers: obsolete

3.

2. Planning only....................changes or uncertainties , but does not eliminate.

0   / 1 pt

Auto-graded

Anticipates	X

Correct answers: anticipate, forecast

The above examples show that even one-word to one-sentence evaluation is too tricky, but the demand for online subjective examination increases. In general, subjective observation examination is considered very important at many levels of the education system, such as primary, secondary, higher secondary, graduation, post-graduation, etc., even in today's scenario. However, it is very challenging to evaluate long answers as they are of varying lengths, paraphrased, and may contain a different type of example that does not match the model answers but is correct.

10.3 PROPOSED MODEL

Figure 10.2 gives a quick overview of the subjective question-answering system.

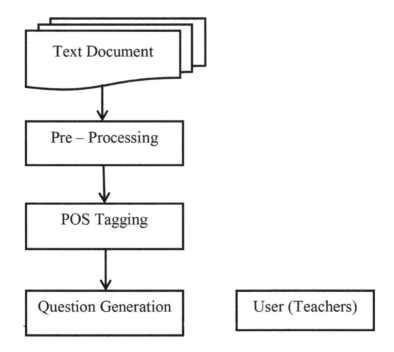

Figure 10.2 Subjective question-answering system.

10.3.1 Text document

Data is passed in a document or paragraph to the system, which can be fetched from the web or any other file (Text file).

10.3.2 Preprocessing

In preprocessing tokenization, POS tagging, Stop Words, and operations for input text are generally performed.

In preprocessing, the following processes are done:

Word-level segmentation.
Sentence-level segmentation.
Punctuation removal.

10.3.3 POS tagging

POS tagging is the process of grouping words into POS and defining them accordingly. Words can be classified as nouns, verbs, adjectives, and adverbs. These classes are known as POS. POS are assigned short tags, such as NN, VB, etc.

10.3.4 Question generation

Question generation is done in two ways:

   i. Manual question generation:
      This question is supplied by the user (Teacher/Admin) based on the input paragraph.
   ii. Auto question generation:
      These questions are generated based on the POS tag for each sentence in the text. This generated question required validation from the user for further processing.

10.3.5 User (students) and user (teacher)

User or student generating the question for the system.

10.3.6 Model answer

A question is generated for every possible sentence, and a model answer is also generated for different assessment purposes to the input question system-generated model answers.

10.3.7 Answer

In this phase, the model answer and question generation answer are calculated for evaluation.

10.3.8 Evaluation

Evaluating the answer from the model answer through the system for finding the appropriate answer with the line numbers of the particular answer, this system does the same procedure for finding the appropriate answer; the only difference is that it contains line numbers at the time of fetching the Knowledge Base (KB) files.

10.3.9 Result

In precision, correct and incorrect answers are calculated, while Recall measures correct and missed values are calculated. Missed or blank values mean the part where the candidate or system has left the question or their answer blank.

10.4 CONCLUSION

The QA system can be of great help in our education system, as in this global era, the need for a QA system that can solve closed domain problems and give a relevant answer to the question has become an essential requirement. Such systems will auto-evaluate or grade long answers with results consistent with human performance. A new possibility of online learning has prompted the significance of online subjective examination conduct and evaluation. Knowledge representation, precise representation for proper understanding, paraphrasing, conceptual learning, online accessing of descriptive questions, evaluation of answers, and inclusion of figures, tables, and mathematical equations are a few significant challenges in front of the researcher. As the aspects involved are complex, their solutions are even more difficult, but the application needs are high. Thus, there is great potential for exploring the challenges in the QA domain.

REFERENCES

Agarwal, M. (2012). Cloze and open cloze question generation systems and their evaluation guidelines. Master's thesis. International Institute of Information Technology, Hyderabad.

Allam, A. M. N., & Haggag, M. H. (2012). The question answering systems: A survey. *International Journal of Research and Reviews in Information Sciences (IJRRIS)*, 2(3), 1–12.

Bin, L., Jun, L., Jian-Min, Y., & Qiao-Ming, Z. (2008, December). Automated essay scoring using the KNN algorithm. In *2008 International Conference on Computer Science and Software Engineering* (Vol. 1, pp. 735–738). IEEE.

Burrows, S., Gurevych, I., & Stein, B. (2015). The eras and trends of automatic short answer grading. *International Journal of Artificial Intelligence in Education*, 25(1), 60–117.

Carrino, C. P., Costa-jussà, M. R., & Fonollosa, J. A. (2019). Automatic spanish translation of the squad dataset for multilingual question answering. *arXiv preprint arXiv:1912.05200*.

Chen, C. Y., Liou, H. C., & Chang, J. S. (2006, July). Fast–An automatic generation system for grammar tests. In *Proceedings of the COLING/ACL 2006 Interactive Presentation Sessions* (pp. 1–4).

Chu-Carroll, J., Ferrucci, D. A., Prager, J. M., & Welty, C. A. (2003). Hybridization in question answering systems. *New Directions in Question Answering, 3*, 116–121.

Coniam, D. (1997). A preliminary inquiry into using corpus word frequency data in the automatic generation of English language cloze tests. *Calico Journal, 14*(2), 15–33.

Das, D. (2008). *Development of dynamic model for substrate uptake and metabolism in microbial and animal cells* (Doctoral dissertation, Indian Institute of Technology, Bombay (India)).

Das, B., & Majumder, M. (2017). Factual open cloze question generation for assessment of learner's knowledge. *International Journal of Educational Technology in Higher Education, 14*(1), 1–12.

Deena, G., Raja, K., Nizar Banu, P. K., & Kannan, K. (2020). Developing the assessment questions automatically to determine the cognitive level of the E-learner using NLP techniques. *International Journal of Service Science, Management, Engineering, and Technology (IJSSMET), 11*(2), 95–110.

Dornescu, I. (2012). Encyclopedic Question Answering (Doctoral Dissertation), retrieved from to the University of Wolverhampton.

Dwivedi, S. K., & Singh, V. (2013). Research and reviews in question answering system. *Procedia Technology, 10*, 417–424.

Fahmi, I. (2009). *Automatic term and relation extraction for medical question answering system.* University Library Groningen][Host].

Girkar, A., Mohitkhambayat, Waghmare, A., Chaudhary, S. (2021). Subjective answer evaluation using natural language processing and machine learning. *International Research Journal of Engineering and Technology (IRJET), 8*(4), e-ISSN: 2395-0056.

Gite, H. R., Asmita, D., & Mahender, C. N. (2013). Development of inference engine to automate the descriptive examination system. *International Journal of Computer Applications, 65*(22), 40–43.

Gupta, D., Ekbal, A., & Bhattacharyya, P. (2019). A deep neural network framework for English Hindi question answering. *ACM Transactions on Asian and Low-Resource Language Information Processing (TALLIP), 19*(2), 1–22.

Harabagiu, S. M., Moldovan, D. I., Pasca, M., Mihalcea, R., Surdeanu, M., Bunescu, R. C., ... & Morarescu, P. (2000, November). FALCON: Boosting knowledge for answer engines. In *TREC* (Vol. 9, pp. 479–488).

Higashinaka, R., & Isozaki, H. (2008). Corpus-based question answering for why-questions. In *Proceedings of the Third International Joint Conference on Natural Language Processing: Volume-I.*

Horowitz, E. (2012). *Fundamentals of programming languages.* Springer Science & Business Media.

Indurkhya, N., & Damerau, F. J. (2011). Handbook of natural language processing. *Computational Linguistics, 37*(2), 395–397.

Jurafsky, D., & Martin, J. H. (2015). Question answering. *Computational Linguistics and Speech Recognition. Speech and Language Processing. Colorado.*

Kakkonen, T., Myller, N., Sutinen, E., & Timonen, J. (2008). Comparison of dimension reduction methods for automated essay grading. *Journal of Educational Technology & Society, 11*(3), 275–288.

Kanaan, G., Hammouri, A., Al-Shalabi, R., & Swalha, M. (2009). A new question answering system for the Arabic language. *American Journal of Applied Sciences, 6*(4), 797.

Khillare, S. A., Shelke, B. A., & Mahender, C. N. (2014). Comparative study on question answering systems and techniques. *International Journal of Advanced Research in Computer Science and Software Engineering, 4*(11), 775–778.

Kolomiyets, O., & Moens, M.-F. (2011). A survey on question answering technology from an information retrieval perspective. *Information Sciences, 181*(24), 5412–5434.

Lopez, V., Uren, V., Sabou, M., & Motta, E. (2011). Is question answering fit for the semantic web? A survey. *Semantic Web, 2*(2), 125–155.

Magnini, B., Romagnoli, S., Vallin, A., Herrera, J., Penas, A., Peinado, V., ... & Rijke, M. D. (2003, August). The multiple language question answering track at CLEF 2003. In *Workshop of the Cross-Language Evaluation Forum for European Languages* (pp. 471–486). Springer, Berlin, Heidelberg.

Mayeesha, T., Md Sarwar, A., & Rahman, R. M. (2021). Deep learning based question answering system in Bengali. *Journal of Information and Telecommunication, 5*(2), 145–178.

Mishra, A., & Jain, S. K. (2016). A survey on question answering systems with classi-fication. *Journal of King Saud University-Computer and Information Sciences*, 28(3), 345–361.

Pino, J., & Eskenazi, M. (2009, March). Measuring hint level in open cloze ques-tions.In *Twenty-Second International FLAIRS Conference*.

Plamondon, L., & Foster, G. (2003, August). Quantum, a French/English cross-language question answering system. In *Workshop of the Cross-Language Evaluation Forum for European Languages* (pp. 549–558). Springer, Berlin, Heidelberg.

Pomerantz, J. (2005). A linguistic analysis of question taxonomies. *Journal of the American Society for Information Science and Technology*, 56(7), 715–728.

Rozali, D. S., Hassan, M. F., & Zamin, N. (2010, June). A survey on adaptive quali-tative assessment and dynamic questions generation approaches. In *2010 International Symposium on Information Technology* (Vol. 3, pp. 1479–1484). IEEE.

Sasikumar, U., & Sindhu, L. (2014). A survey of natural language question answering system. *International Journal of Computer Applications*, 108(15), 975–8887.

Seena, I. T., Sini, G. M., & Binu, R. (2016). Malayalam question answering system. *Procedia Technology*, 24, 1388–1392.

Thaker, R., & Goel, A. (2015). Domain specific ontology based query processing system for Urdu language. *International Journal of Computer Applications*, 121(13), 20–23.

The START Natural Language Question Answering System. (2015). Retrieved September 04, 2015, from http://start.csail.mit.edu/index.php.

Trigui, O., Belguith, L. H., & Rosso, P. (2010). DefArabicQA: Arabic definition question answering system. In *Workshop on Language Resources and Human Language Technologies for Semitic Languages, 7th LREC, Valletta, Malta* (pp. 40–45).

Vignesh, N., & Sowmya, S. (2013). Automatic question generator in Tamil. *International Journal of Engineering Research & Technology*, 2(10), 1051–1055.